THE
EARTH

A Visual Story of Our Amazing Planet
Featuring NASA Images

BETH ALESSE

T0178847

AMHERST MEDIA, INC. ■ BUFFALO, NY

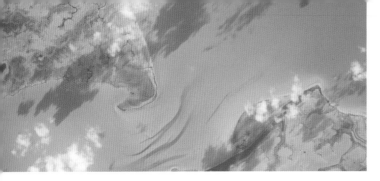

Beth Alesse is a graphic artist, editor, and author. She holds degrees in art and education and has a background in graphic arts, linguistics, and visual and audio digital media. Her curated image collections have appeared in media and books, some of which she has written.

Published by:
Amherst Media, Inc., P.O. Box 538, Buffalo, N.Y. 14213
www.AmherstMedia.com

Publisher: Craig Alesse
Senior Editor/Production Manager: Michelle Perkins
Editors: Barbara A. Lynch-Johnt and Beth Alesse
Acquisitions Editor: Harvey Goldstein
Associate Publisher: Katie Kiss
Editorial Assistance from: Carey Miller, Ray Bakos, Jen Sexton-Riley, Rebecca Rudell
Business Manager: Sarah Loder
Marketing Associate: Tonya Flickinger

ISBN-13: 978-1-68203-316-6
Library of Congress Control Number: 2017949335
Printed in The United States of America.
10 9 8 7 6 5 4 3 2 1

www.facebook.com/AmherstMediaInc
www.youtube.com/AmherstMedia
www.twitter.com/AmherstMedia

AUTHOR A BOOK WITH AMHERST MEDIA

Are you an accomplished photographer with devoted fans? Consider authoring a book with us and share your quality images and wisdom with your fans. It's a great way to build your business and brand through a high-quality, full-color printed book sold worldwide. Our experienced team makes it easy and rewarding for each book sold—no cost to you. E-mail **submissions@amherstmedia.com** today

Contents

Earth Images From Space

This book, which is part of a trilogy: *The Earth, The Sun, and The Moon,* is full of beautiful and awe-inspiring images. Some are taken by astronauts, and some are made from satellite data. Many are recognizable, often looking like the maps that have been drawn of the same areas. Other images appear like abstract paintings with colors and textures that show the grandeur of the surface. They offer us a new vantage point, a perspective that is relatively new to humans, of the places where we live, of our home planet Earth. Whether recognizable or not, the artistic and technical hands that have created these images have an eye for design and aesthetics. These images are a joy and an inspiration. They provide access to exploration and understanding of Earth.

The National Aeronautics and Space Administration (NASA) and its international communities share these Earth images with everyone, and the data collected, with the world. Imaging techniques, and data from satellite instruments are continually being refined and updated. It is used worldwide to improve crops, fight fires, aid in emergency response and disaster relief, such as landslides and floods, predict weather, track drought conditions, monitor snow coverage, and identify climate variances.

As the editor/curator, I have tried to give image credit as requested by the websites where each image was acquired. Similar images were available from different websites with variations in credits provided. If I have left anyone out, please contact me and corrections will be made in future editions. Also, if you would like to explain your imaging process, feel free to reach out. For those whose images were not included, I would like to hear from you, too—indeed, choices were difficult to make.

Images and data collection are advancing, fostering new and significant discoveries about our Earth. These will be epoch-making discoveries that will enlighten the caretakers of our planet. Enjoy this collection of amazing images of Earth.

Beth Alesse
BAlesse@AmherstMedia.com

Early Film Space Images

In 1609, Italian scientist Galileo Galilei used the newly invented telescope to view the night sky. Galileo viewed the moon, planets, stars, and beyond. His observations conclusively changed humankind's grasp of the universe. Nearly four centuries later, NASA and its associates have turned the table and are now viewing Earth from space. The images from cameras, and the data from satellite instruments, are changing our understanding of the Earth.

Image credit *(facing page and bottom):* NASA

This image of the Earth *(facing page)* was taken by the crew of the Apollo 11 during their outbound trip from the Earth to the Moon. The image below was taken on the same mission on its homeward journey.

Astronauts Neil A. Armstrong, Michael Collins, and Edwin E. Aldrin Jr., were on board the Apollo 11 spacecraft with the famed Hasselblad camera. Most of these early images were taken with the Hasselblad camera using film and a 70mm lens.

First American to Walk in Space

The first American to walk in space was Edward H White II in 1965, on the Gemini 4 mission. The device that he has in his right hand is not a camera but a Hand Held Self Maneuvering Unit (HHSMU). The unit helped him move about in the weightlessness of space. The 23-foot tether line anchored him to the Gemini spacecraft.

Astronaut James McDivitt was inside monitoring the space walk and communications. McDivitt was also responsible for taking amazing, well-shared, and historic photographs of the walk.

Image credit : NASA

Image credit : NASA

Gemini 7, 1965

This image of the Andes Mountains with scalloped waves of clouds, was taken from the Gemini 7 space-craft in 1965.

The Earth's terminator can be seen on the left where there is a dividing line between the light and dark on the planet, or in other words, the line between night and day.

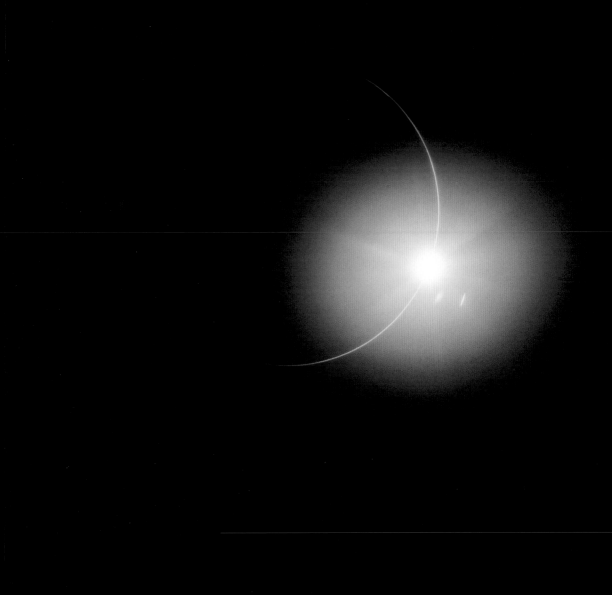

Apollo 12, 1969

This amazing view was photographed from the Apollo 12 spacecraft on its journey back from the Moon. The Earth moved between the Apollo 12 and the Sun. It is a unique eclipse that had never been seen before, especially from the planet Earth.

Skylab 4, 1974

Skylab space station photographed
these ice formations of Hudson Bay,
Canada. Like the Apollo and Gemini
missions before, a 70mm Hasselblad
with film, was used for photograph-
ing the Earth.

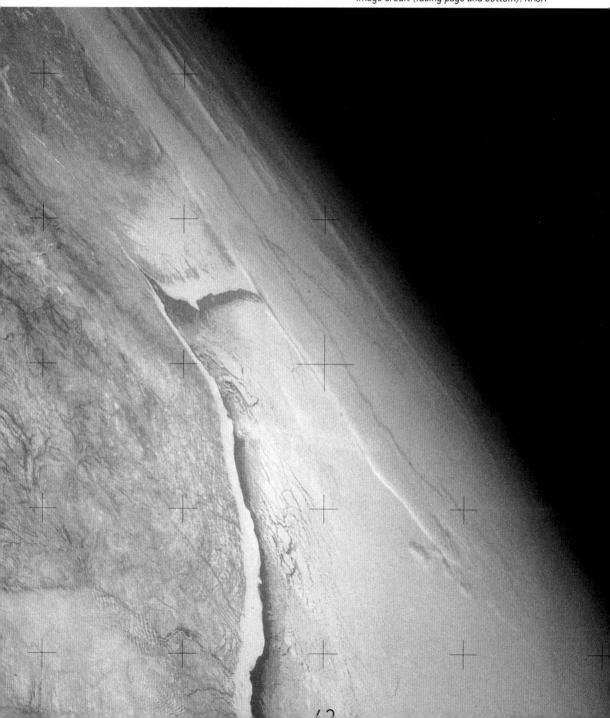

Image credit: NASA

STS-64,
Space Shuttle
Discovery, 1994

The Shuttle was 130
nautical miles above the
Earth when this image
was taken with a handheld
Hasselblad camera.

Seeing Earth

The Apollo missions used film cameras. Between the years 1981 and 2011, many images came from NASA's space shuttles. Today we see images of Earth mostly from digital still and digital motion cameras aboard the International Space Station (ISS), and from instruments on artificial satellites that collect data as they orbit the Earth. This image was taken in 2016 from aboard the International Space Station in the Cupola module windows using a Nikon D4 camera.

Cameras in Space

John Glenn was the third American in space and the fifth person ever in space. Glenn flew the Friendship 7, a Mercury spacecraft, on February 20, 1962. He decided to bring along a modified Ansco Autoset 35mm camera, made by Minolta. This image *(bottom)* is one of the first photographs of the Earth from space.

On later missions the Hasselblad *(facing page, bottom)* was used extensively by NASA for many years and missions, until lighter options became available. Often the photographs were taken with a handheld camera. Sometimes the cameras were modified, for example, anchors or attachments were added allowing them to be attached to spacesuits. Hasselblad cameras were also taken to the Moon, and photographs of the Earth were taken in transit.

Infrared, ultraviolet, motion, and other specialized cameras have been used on various missions. More recently, Nikon, Sony, and Kodak digital still and video cameras have been used. NASA has also developed the High Definition Earth-Viewing System (HDEV), and has experimented with Synchronized Position Hold Engage and Reorient Experimental Satellites (SPHERES). SPHERES are intended to be operated semi-autonomously.

Resolution has improved considerably and continues to be refined. Likewise, lens technology has had improvements, too.

Image credit: NASA, John Glenn astronaut photographer

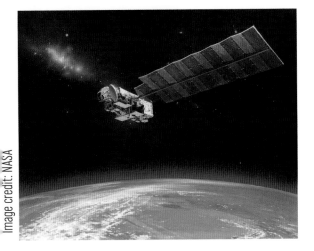

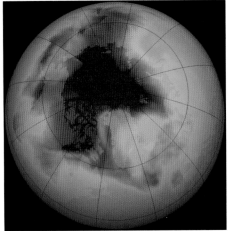

Satellites and Instruments

Satellites that orbit the Earth can have a variety of instruments to create images, which can be used with or without photographs. These images are not only beautiful, they help to visualize the data collected. An example is the image *(top right)* that maps the ozone measured by the Ozone Monitoring Instrument (OMI) on NASA's Aura satellite. The Aura is one of three major components of the EOS (Earth Observing System). The other two components of the EOS are Aqua and Terra. The Aura satellite also holds these instruments: High Resolution Dynamics Limb Sounder (HIRDLS), Microwave Limb Sounder (MLS), and Tropospheric Emission Spectrometer (TES).

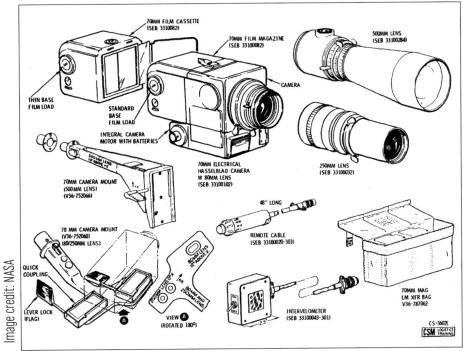

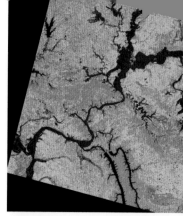
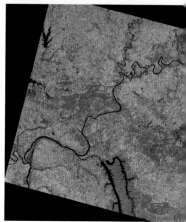

Image credit: NASA

Satellites and Data

It has been estimated that there may be as many as 3,600 satellites in Earth's orbit, 1,000 of which are operational. Not all are Earth observation satellites. Some are for communications, navigation, and space telescopes. Some are in low Earth orbit, polar orbit, and geostationary orbit.

Upgrades and international partnerships are an important part of satellite operations. For example, the Gravity Recovery and Climate Experiment (GRACE) satellite was a joint mission, a partnership between Jet Propulsion Laboratory (JPL) and the German Research Centre for

Geosciences (GFZ). It is updated with the GRACE-FO satellite. Another satellite, Land Remote-Sensing Satellite system (Landsat), is a series of nine satellites. The Landsat 8 is a partnership between NASA and the United States Geological Survey (USGS). Partnerships can be specific to individual instruments on board a satellite.

Images are made from combining data from a variety of instruments, time periods, photographs, and maps. These images are of the confluence of the Wabash and Ohio Rivers. The image of normal flow *(bottom right)* was combined with an image from spring flooding *(top right)* to make the final image *(facing page, left)*. False color bands used in this image better distinguish the flooded areas.

NOAA's GOES-16 satellite is used for weather forecasting, distinguishing between kinds of water vapor, smoke, ice, and ash. It uses the moon for calibration. This satellite from NASA Worldview *(bottom)* shows Puerto Rico in the throes of Hurricane Maria.

The Experimental Advanced Airborne Research Lidar (EAARL), a collaboration of NASA and USGS, is used on aircraft to take laser readings

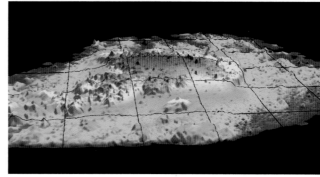

Image credit *(top):* John Brock of the United States Geological Survey and Wayne Wright of NASA's Wallops Flight Facility

Image credit *(bottom):* NASA Worldview

to make topographic maps of surfaces *(top)*, especially those below water.

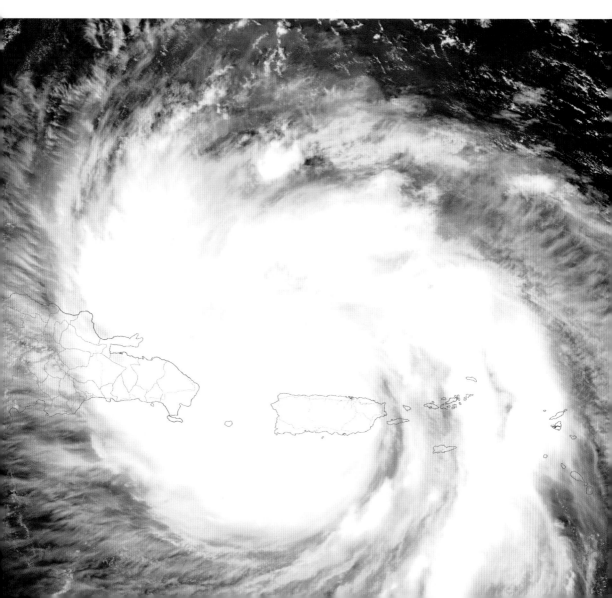

Image credit: NASA Earth Observatory image by Jesse Allen, using Landsat data from the U.S. Geological Survey

Rapid Response System

Initially, the Rapid Response system used the MODIS instrument on the Terra Satellite to help the U.S. Forest Service (USFS) and the National Interagency Fire Center (NIFC). Soon other federal, state, and media users made requests for these real-time images.

This image *(facing page)* and others like it help firefighters and rescue workers with real-time imagery. Unlike true-color photographs, non-visible information such as infrared, heat, or ultraviolet can be assigned colors, making that information *visible* in the image. False-color images showed the movement of the King Fire in California, 2014, and gave firefighters more timely and accurate information about the fires they were fighting.

Thermal Images

Thermal images inform meteorologists about what is happening inside the storm. The top image is a representation of Hurricane Noel in 2007.

The thermal image of Hurricane Maria *(bottom)* on September 20, 2017, in about the same position, over Puerto Rico, as seen on page 21. Thermal images help to assess the power of the thunderstorms in Maria's eyewall. The Visible Infrared Imaging Radiometer Suite (VIIRS) instrument on NASA-NOAA's Suomi NPP satellite was used to capture this image *(bottom)*.

Image credit *(top):* NASA Goddard Rapid Response Team

Image credit *(bottom):* NOAA

Much of what we see in Earth images taken from space is recognizable. However, so much of what we see in these images is new and foreign to our eyes. It's like a microscope or a macro lens. Images from space are informative and give an enlightened shift in views of Earth.

A Lava Field

This is a lava field *(facing page)* in southern Syria. Looking like an abstract painting, it has the linear rows of circles (most prominent in the center and upper left of the image), which are chains of cinder cones from past volcanic activities. The Advanced Spaceborne Thermal Emission and Reflection Radiometer (ASTER), a Japanese instrument on NASA's Terra satellite, was used to capture this image in 2009.

Images Communicate

This image *(bottom)* is interesting and colorful, and not exactly real looking. Some would say it's beautiful. But most importantly, it communicates scientific findings, almost like a graph or chart. The color information is from radar data. The concentric circles in the middle represent the sinking of the ground's surface adjacent to the Calbuco Volcano, Chile, due to the decreased magma below, which was the result of an eruption.

Image credits *(facing page):* NASA/GSFC/METI/ERSDAC/JAROS, and U.S./Japan ASTER Science Team

Image credits *(bottom):* ESA/NASA/JPL-Caltech

Beautiful Earth

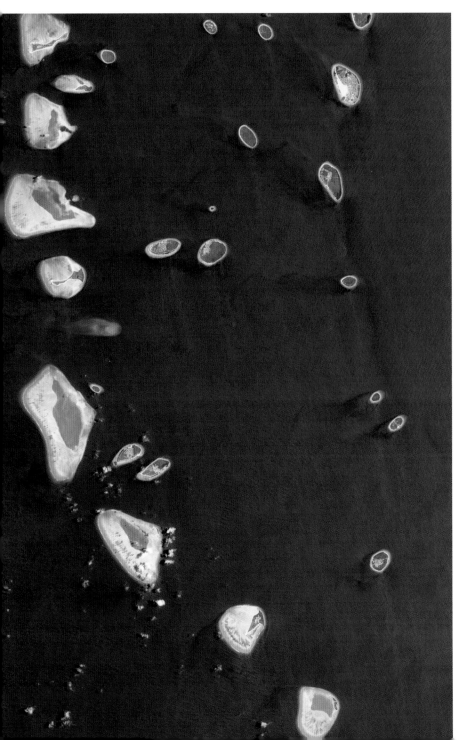

The images of earth taken by astronauts, satellite instruments, and space telescopes are exquisite with patterns, details, lines, shapes, and colors. The scientists, technicians, and artists who work with data to compose these collections are not only advancing science, and stewardship of our world; they bring to all of us the shared beauty of our Earth.

Lowest Country

The Republic of Maldives is the lowest country in the world with an average elevation of 4 feet, 11 inches (1.5

Image credits: NASA/GSFC/METI/ERSDAC/JAROS, and U.S./Japan ASTER Science Team

Image credit: NASA/METI/AIST/Japan Space Systems, and U.S./Japan ASTER Science Team

meters) above sea level. This makes the country vulnerable to storms and climate change.

Cancun

Cancun, Mexico *(above)*, is a resort city imaged by the ASTER instrument on the Terra satellite in 2014.

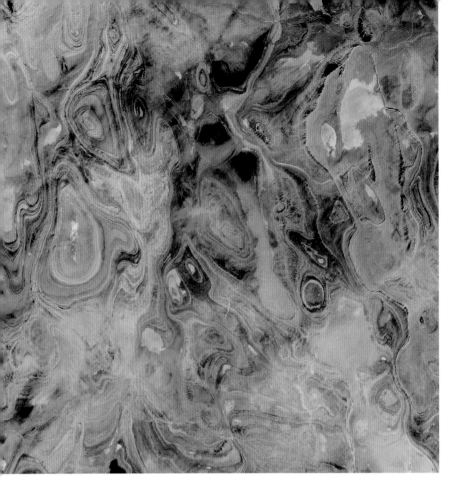

A Dry Basin

This arid place is know as the Tanezrouft Basin, located in the Sahara Desert. The Landsat 8 was used to make this stunning natural-color image.

Ancient Dunes

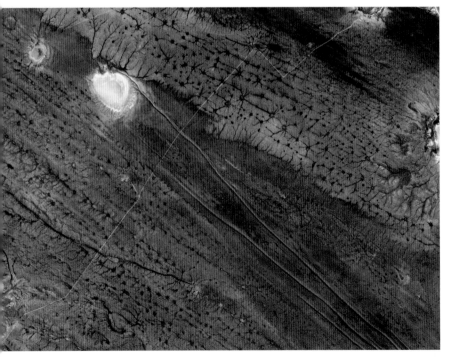

This red image was captured in 2018 by the Copernicus Sentinel-2 mission and processed by ESA. It shows the western part of the Kalahari Desert, where the sand dunes are over 12,000 years old. Roads can be seen cutting through the landscape.

Image credit *(top):* NASA Earth Observatory image by Jesse Allen, using Landsat data from the U.S. Geological Survey

Image credit *(bottom):* ESA

Checkerboard Forest

Taken by an astronaut on the International Space Station in 2017, this checkerboard is actually a forest in Idaho. Younger, shorter trees are snow topped and appear white. The older, taller trees are a dark brown.

Operation IceBridge

NASA's Operation IceBridge helps predict the Arctic Sea ice conditions. It is an airborne survey that maps the extent, frequency, and depth of melt ponds.

Image credit *(top):* NASA

Image Credit *(bottom):* NASA/Operation IceBridge

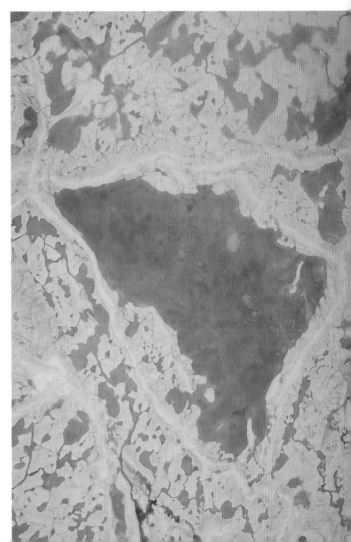

Clouds Over Peru

Clouds collect in the deep valleys of Peru on July 26, 2015 *(top)*. Image was captured with instruments on Landsat 8.

Glacial Drumlins Become Islands

This is an example of sunken glacial drumlins *(bottom)*, and is located in Clew Bay in County Mayo, Republic of Ireland, May 31, 2016. These low hills, now surrounded by water, are sediment deposited by melting glaciers.

Image credit: NASA/METI/AIST/Japan Space Systems, and U.S./Japan ASTER Science Team

First Norse Settlement of Greenland

This image, acquired June 13, 2016, shows the first Norse settlement of Greenland. It is radiocarbon dated to be about the year 1000 AD.

This settlement *(yellow asterisk in the center)* called Brattahlid in English, is located near the present-day settlement of Qassiarsuk. It was the southwestern Greenland estate of Erik the Red, father of Leif Erikson, the famous Icelandic explorer.

The Rocky Mountains

Mountains, by their sheer height, often impede the movement of clouds. This image of the Rocky Mountains was photographed on the International Space Station by Expedition 50 Flight Engineer Thomas Pesquet of the European Space Agency who shared it on social media, January 9, 2017.

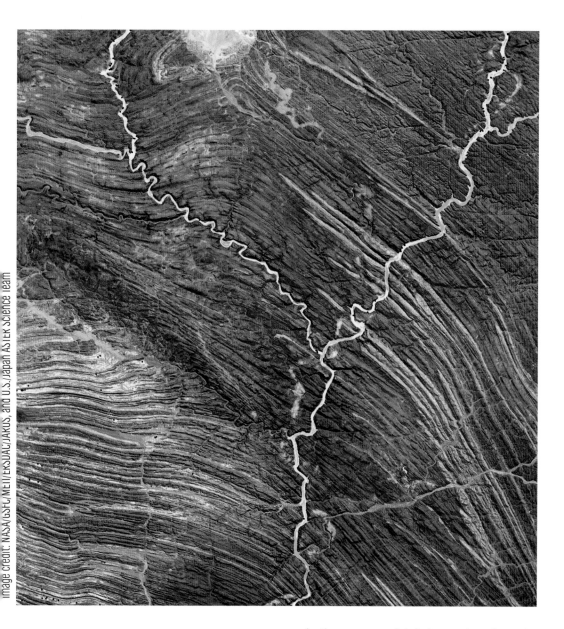

Image credit: NASA/GSFC/METI/ERSDAC/JAROS, and U.S./Japan ASTER Science Team

The Ugab River

This is a false-color image of the Ugab River as it runs through northern Namibia. The image was taken December 25, 2000, by the ASTER instrument on NASA's Terra satellite.

Much of the land in this image is made of sedimentary rock, sandstone and siltstone—which is made of particles that are finer than the sandstone. NASA's Earth Observatory used this image for representing the letter "Y" in its collection of images from NASA satellite imagery and astronaut photography that resemble the 26 letters of the English alphabet.

Image credit *(top and facing page):* NASA

Cape Verde Islands

This photograph *(above)* was taken from the Shuttle Atlantis during the STS-125 mission (May 11 to 24, 2009) to repair and update the Hubble Space Telescope, May 15, 2009.

These low clouds are over the Cape Verde Islands. The official name of the country is Republic of Cabo Verde. It is made up of ten volcanic islands in the Atlantic Ocean off the northwest coast of Africa. This archipelago is volcanic in origin. Pico do Fogo is the country's largest active volcano, last erupting in 2014.

These two images *(facing page)* show the shuttle's Canadian-built remote manipulator system used in the servicing mission. Mission Specialists John Grunsfeld (at the end of the manipulator system) and Andrew Feustel *(top center)* participate in the final space walk of the mission working high above the Earth.

Florida Straits

This image *(above)* is taken from Gemini IV, June 4, 1964 with a Hasselblad camera and a 70mm lens. Part of astronauts Jim McDivitt and Ed White's mission was to take photographs of Earth's weather and terrain.

Lake Eyre, Australia

Lake Eyre *(facing page, top)* is an ephemeral salt lake receiving less than an average of 5 inches of rain a year. It is located at a point in Australia that is almost 50 feet below sea level. The size and shape of the lake changes depending on the rainfall and weather.

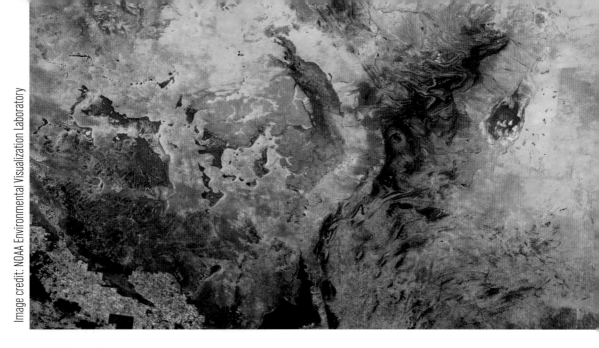

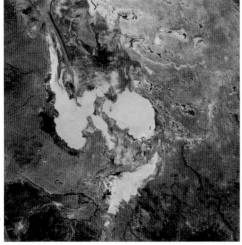

The thermal images taken of the Lake Eyre area vary greatly depending on the season and rainfall. All three images are striking. The top image, taken in 2016, is thermal imagery and uses VIIRS SVI channels. The middle image, taken in 1999, is a composite, using infrared, near infrared, and blue wavelengths from the Landsat 7 satellite. The bottom image was taken by the crew of the Space Shuttle Columbia in 1990.

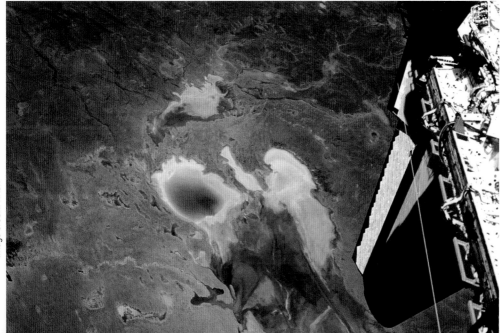

Image credit: NASA

Plankton Bloom

This image *(facing page)* was composed using chlorophyll data and data from the red, green, and blue bands from Visible Infrared Imaging Radiometer Suite (VIIRS) on the Suomi National Polar-orbiting Partnership (Suomi NPP) weather satellite. The processing highlights color differences and bring out some of the subtle features of the phytoplankton bloom.

Many Kinds of Light

An auburn-golden sunset *(top)* was taken from the International Space Station on April 18, 2015. The image is packed with so much: hints of the station's structures, city lights on the ground, lightning in the clouds (in the lower middle and right of the image), and the red light of the aurora above the Earth's horizon.

Flooded Land

This image *(right)* shows Hurricane Harvey's inland flooding in the Houston area, Texas, taken by NASA's Terra satellite on September 5, 2017.

Image credit *(above):* NASA/METI/AIST/ Japan Space Systems, and U.S./Japan ASTER Science Team

Image credit *(facing page):* NASA image by Norman Kuring, using VIIRS data from the Suomi National Polar-orbiting Partnership

Glacier Cracks

In 2013, the Operational Land Imager (OLI) on the Landsat 8 satellite took this image *(above)* of an iceberg separating itself from the edge of Pine Island Glacier.

Image credit *(top):* NASA Earth Observatory images by Holli Riebeek, using Landsat 8 data from the USGS Earth Explorer

Run Island, Indonesia

Run Island, of the Banda Islands, is on the left *(below)*. These islands were originally the only source of nutmeg and mace. The image was taken with ASTER, and acquired January 5, 2016.

Image credits *(bottom):* NASA/GSFC/METI/ERSDAC/JAROS, and U.S./Japan ASTER Science Team

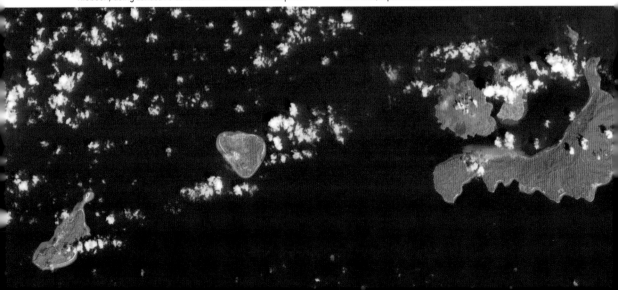

Sahara Desert

The Sahara Desert *(top)* is nearly the size of China or the United States and comprises much of Northern Africa. Landsat 7 took this image of large bands in the Sahara Desert near the Terkezi Oasis in the country of Chad.

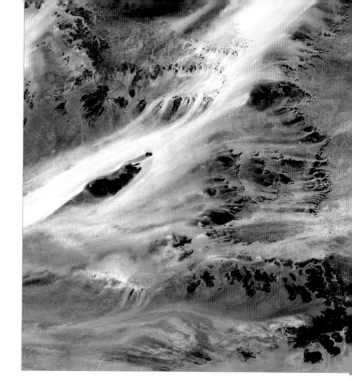

Over 3,000 Glaciers

This image *(middle)* of New Zealand's glaciers was taken with the ASTER instrument on Terra. New Zealand has over 3,000 glaciers, most of which have been retreating for over a century. A few have short periods of advancement, but overall they do not appear to be regaining their magnitude.

Ephemeral Lakes

This image *(bottom)* is of Western Australia's Lake Mackay. Seen here are hundreds of smaller ephemeral lakes. The different colors indicate desert vegetation, algae, water pools, and different degrees of soil moisture.

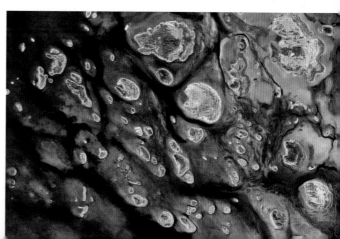

Image credit *(top):* NASA

Image credits *(middle):* NASA/GSFC/METI/ERSDAC/JAROS, and U.S./Japan ASTER Science Team

Image credits *(bottom):* NASA/GSFC/METI/ERSDAC/JAROS, and U.S./Japan ASTER Science Team

The Aral Sea

Once one of the largest lakes in the world with vast wetlands, the Aral Sea *(above)* is drying up. The dark areas indicate where water remains.

Mississippi River Silt

The Mississippi River *(below)* flows through Louisiana and out to the Gulf of Mexico. The shades of blue in this image indicate the silt carried by the river.

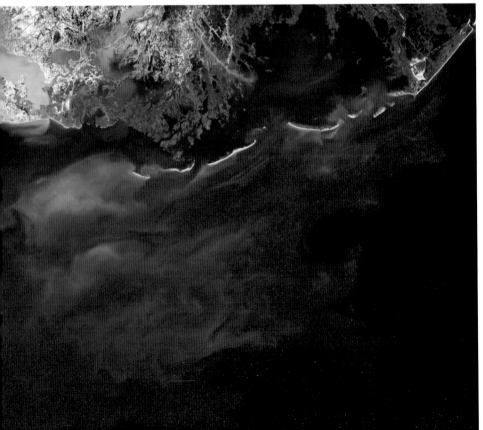

An Astronaut's View

This is what astronauts see looking out of their cupola window *(above)*. The dome affords a view of a portion of the International Space Station and an amazing view of the planet.

False-Color Patterns

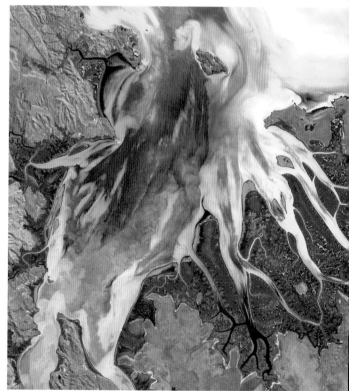

False-color patterns *(right)* are in an estuary image of Western Australia indicating the sediment and nutrients.

Atmosphere

When we look up from the ground, the atmosphere seems endless. But from the International Space Station, we can see the true thickness of the atmosphere on Earth's horizon. This thin covering on our planet protects all life. The atmosphere warms the air, keeps tem-

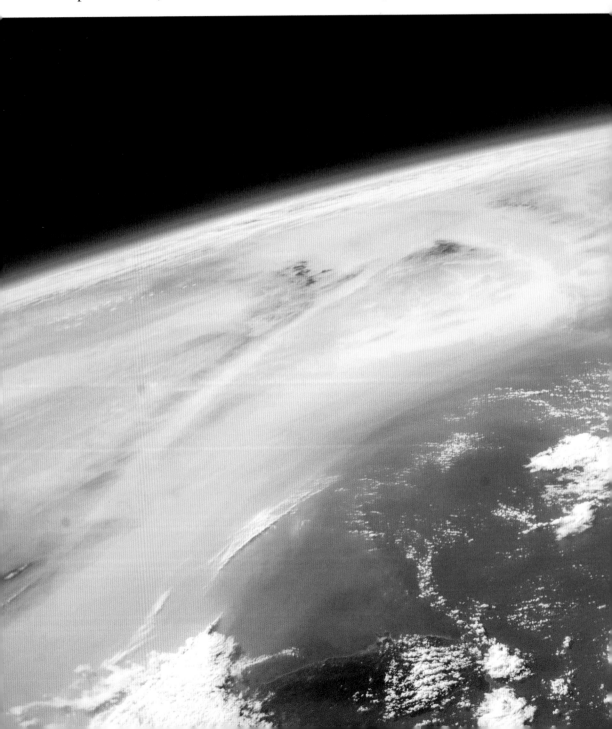

peratures from extreme differences, and keeps water in a liquid state. The atmosphere is where weather happens: the clouds, the rain, wind, snow, and sunshine. Also, the storms happen here, such as hurricanes, thunderstorms, and spring showers.

The water cycle as we know it needs the atmosphere. Rain gives us fresh water for drinking, plant growth, and farming. The atmosphere could never be what it is today except for the effect that living organisms have had on it, starting millions of years ago. Plant life continues to keep our atmosphere healthy.

Image credit: NASA, International Space Station

Weather

Weather is used to describe the precipitation, temperature, wind, humidity, cloud coverage, and other factors in the atmosphere that affect a specific location at a particular time. Weather is variable.

Climate

Climate is used to describe the expected weather that occurs over an extended period of time for a specific place. It is the average weather of a location and time of year.

The Great Lakes

These images of the Great Lakes were created using a combination of infrared and visible light from the Moderate Resolution Imaging Spectroradiometer (MODIS) instruments on the Terra and Aqua satellites. Open water that is clear of ice looks black. Snow cover is light blue. Lake Erie *(lower right and detailed image above)* is the shallowest of all the lakes, and consequently, freezes more readily.

Image credit *(left and top)*: NASA Earth Observatory, Jeff Schmaltz, using MODIS data from LANCE/EOSDIS Rapid

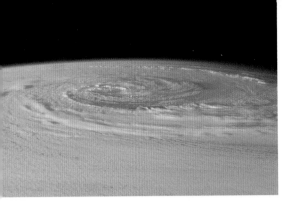

Clouds

Clouds, water vapor in the atmosphere, appear with what seem to be infinite variations in images from space. The clouds are in a huge cyclonic formation *(top)* so thick, nothing can be seen below them.

The clouds above the Philippine Sea *(middle)* show their shadows on the water below. These clouds over Eastern Russia *(bottom)* are aligned with wind. The cloud images to the right, as part of the ISS Science for Everyone program, used Windows on Earth as a tool.

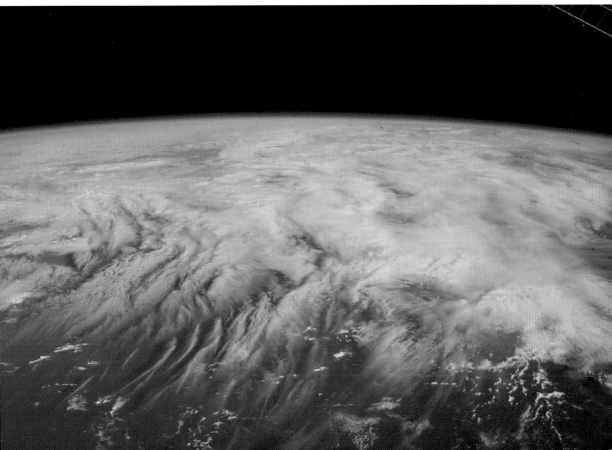

Sprites

Earth's atmosphere is relatively thin, as can be seen in this night time image. Its comparable size can be seen through the transparent horizon. The atmospheric layer has been transformed by life on Earth. It also protects Earth's abundant and diverse life. Many systems and phenomena are a part of the atmosphere, and we are still learning about them.

One fairly recent discovery is the occurrence of sprites, first photographed in 1989 and seen in these images as pinky-red light. Sprites are an electrical discharge of cold

plasma that happens during thunderstorms. Unlike lightning, sprites are cold discharges.

In the image below, the large light above the horizon is the rising Moon over Texas. The yellow lights are from Houston, Dallas, and their surrounding communities.

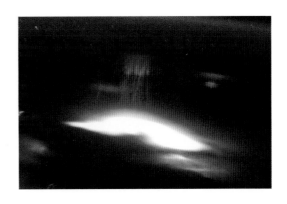

Image credits: NASA/JSC Gateway to Astronaut Photography of Earth

Haze, Pollution, and Dust

Korea U.S.-Air Quality, (KORUS-AQ) is a joint effort of NASA and the Republic of Korea to study and monitor the air quality over South Korea. This 2007 satellite image *(below)* shows a patch of air pollution as it travels across the Korean peninsula and heads toward Japan.

The Taklamakan Desert *(facing page, top)* is in northwest China. Surrounded by mountains on three sides, wind blows dust and sand causing dunes and haze to form and shift. The Suomi NPP VIIRS instrument was used to make this image.

The haze in this image *(facing page, bottom)* is caused by air pollution. According to the United Nations, pollution causes 2 million deaths per year in the Western Pacific.

Image credit: NASA

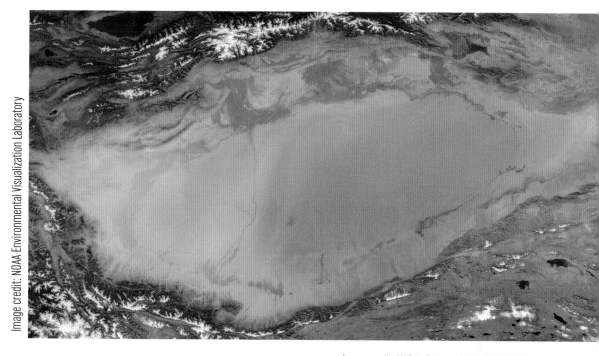

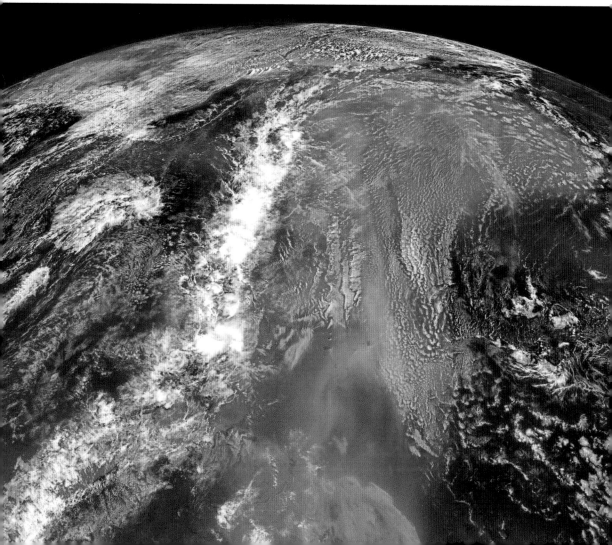

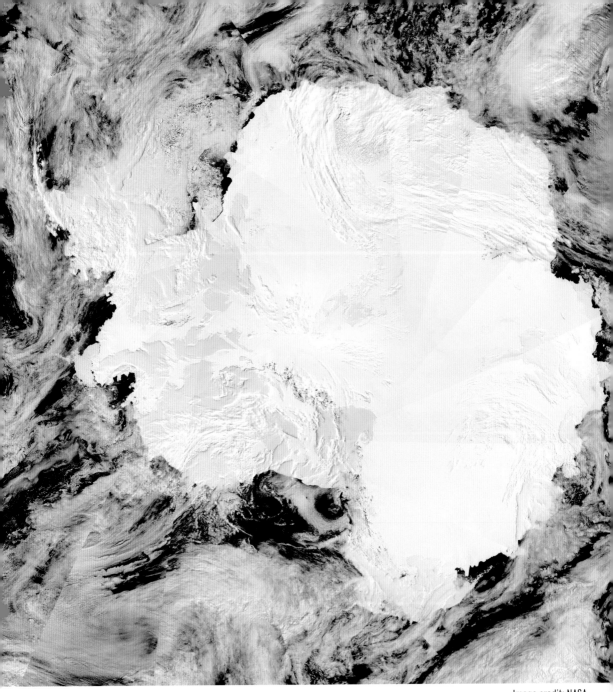

Coldest Place on Earth

The East Antarctic Plateau, Antarctica *(above)* holds the record for the coldest place on Earth: −136F (−93.2C) set Aug. 10, 2010.

Northeastern Siberia *(facing page)* is the coldest, continuously inhabited place on Earth, reaching temperatures of −90F (−67.8C).

Image credit: NASA image by Jeff Schmaltz, LANCE/EOSDIS Rapid Responser

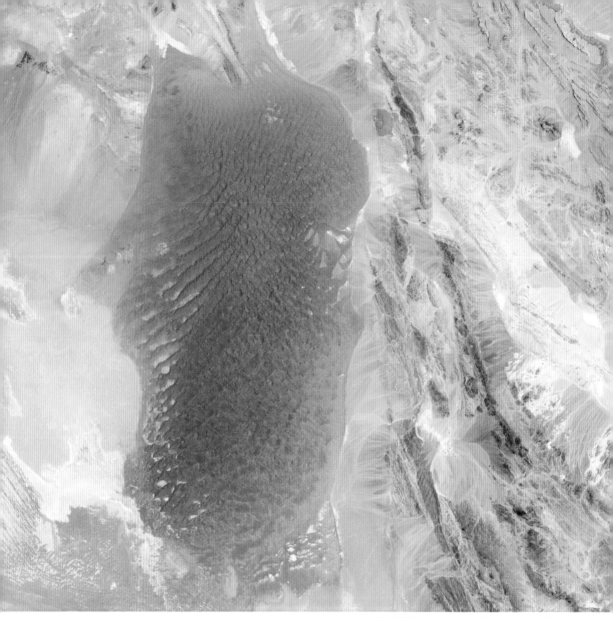

Hottest Place on Earth

The record for the hottest place on Earth is El Azizia, Libya, at 136.4F (58.0C) in 1922. The runner-up (and previous record holder) is Death Valley, California, at 134F (56.7C) in July 1913.

Previous records relied on land-based instruments that measure the air above; today, satellites can measure the actual land temperature. As a result of measuring the surface over a several-year period, the Lut Desert had the highest single temperature recorded in 2005 at 159.3F (70.7C).

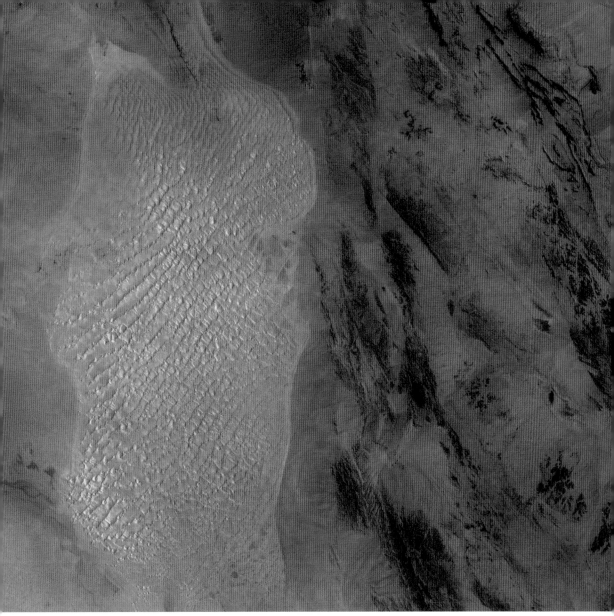

Image credits: NASA images by Jesse Allen and Robert Simmon,
using Landsat 7 data from the USGS Global Visualization Viewer

The image on the facing page is the natural color. The image above shows the temperature of the land. The dark areas are cooler. The lighter areas are hotter.

Near to these record temperatures were the shrub lands of Queensland, Australia at 156.7F (69.3C) in 2003, and the Flaming Mountain in China at 152.2F (66.8C) in 2008.

Image credit: NASA

Driest Place on Earth

Records indicate that Atacama Desert is the driest place on Earth. Although very unusual for this area, the top images on this page show when it received 32 inches of snow in July 2011. These images are from the Moderate Resolution Imaging Spectroradiometer (MODIS) on NASA's Terra satellite. On the left *(above)* is a natural color image. On the right *(above)*, the image includes infrared, too. The dark red is snow. The light red and white areas are clouds. More typical of this desert is the image below that shows mining within the Atacama Desert.

Wettest Places on Earth

The winner for the wettest place on Earth is Mawsynram, Meghalaya State, India *(right)*, which has an average rainfall of about 467 inches (11.87m) It had a record 1000 inches (25m) of rain in 1985.

Although Maui, Hawaii *(below)* is only in seventh place, it nonetheless receives an average of 404 inches (10.27m) of rain a year. As seen in this image, the Hawaiian islands are often shrouded in clouds or "vog." Vog, similar to smog, is produced when sulfur dioxide plus other gases and particles from an erupting volcano react with oxygen and moisture in sunlight.

Image credits *(top):* NASA/GSFC/METI/ERSDAC/JAROS, and U.S./Japan ASTER Science Team

Image credits *(bottom):* NASA/GSFC/METI/ERSDAC/JAROS, and U.S./Japan ASTER Science Team

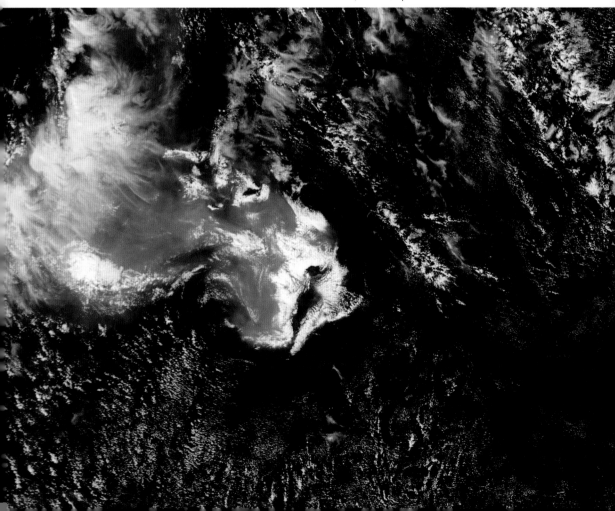

Water

ater is what makes the planet Earth unique in our solar system and beyond. Every kind of life depends on its presence. As a liquid, gas, or solid, water is part of the Earth's atmosphere, oceans, land, and crust. It also affects the Earth's weather and climate systems as it passes through the different phases of the water cycle and energy is exchanged. Water can also transport minerals and organisms.

NASA/NOAA Partnership

This image was acquired using the Visible Infrared Imaging Radiometer Suite (VIIRS) sensor on the Suomi National Polar-orbiting Partnership (Suomi NPP) weather satellite on April 9, 2015.

Image credit: Ocean Biology Processing Group at NASA's Goddard Space Flight Center

Swirls of Sea Ice

Swirls in the sea water *(top)* off the east coast of Greenland are white because they are full of sea ice.

Danny

Hurricane Danny, August 2015 *(bottom)*, was photographed by astronaut Scott Kelly from the International Space Station.

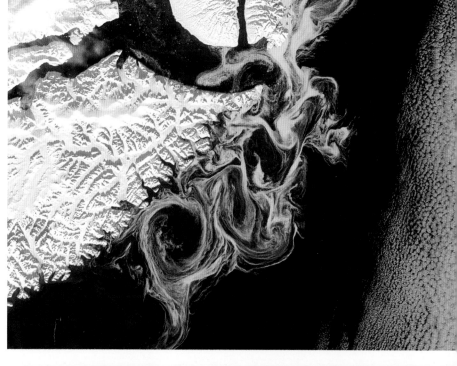

Image credit *(top):* courtesy Jeff Schmaltz, LANCE MODIS Rapid Response Team at NASA GSFC

Image credit *(bottom):* NASA

So Much Water

Water covers about 75 percent of the Earth's surface in the form of salty oceans, fresh water lakes, rivers, swamps, reservoirs, canals, frozen ice caps, and glaciers. Water can be found in the Earth's crust and is the fluid in every living organism. Scientists

speculate that large amounts of water extend deep below the crust, too. It is in all the clouds that are in the sky, and in the atmosphere even when there are no clouds, as humidity. Water is one of the necessary things that make our planet habitable for every form of life. The amount and quality of water substantially determines each ecosystem found on the Earth.

Image credit : NASA

So Little Water

The Earth is 7,917.5 miles across, from one side, through its center to the other side. If all the water on the Earth fit into its own sphere, it would be 860 miles across.

The blue water sphere *(below)*, placed above and west of the North American continent, illustrates the total amount of water on Earth compared to the total size of the planet.

It seems small compared to the apparent size of the oceans. However, the oceans are relatively shallow compared to the volume of the rest of the planet.

Earth's water is continuously transformed as it cycles through the atmosphere, the planet's surface, the ground, the oceans, and living organisms. These systems are complex—but also fragile and tenuous, impacted by both natural disasters and the footprint of human life.

Image credit *(bottom and facing page):* NASA

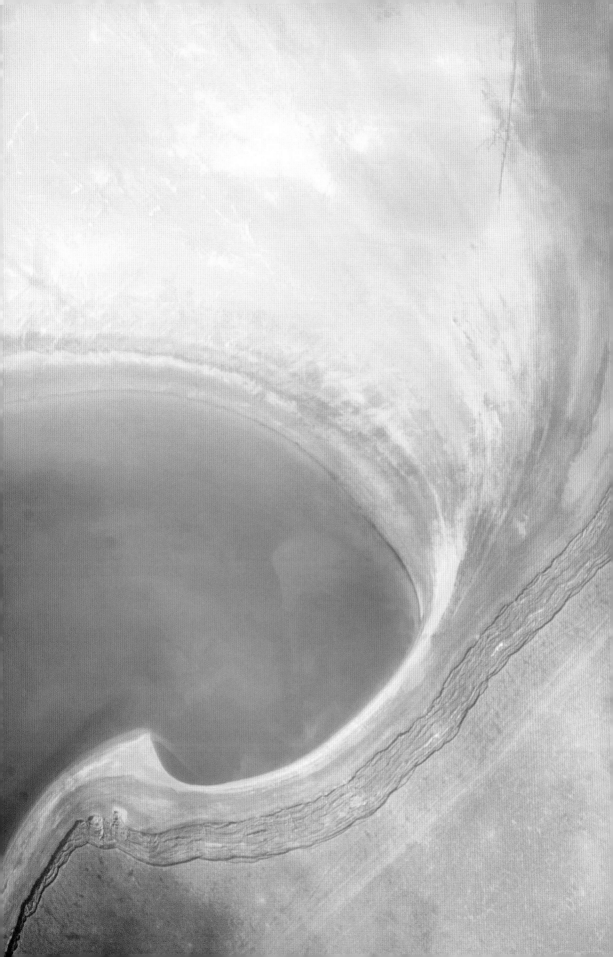

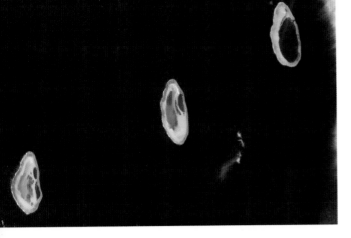

Earth's Oceans

The Earth's saltwater oceans hold 97 percent of all the water on our planet. The average depth of these bodies is nearly 12,100 feet (3,700 meters), which makes it difficult for humans to visit most of the oceans' floors. As a result, it is estimated that only 5 percent of the world's oceans have been explored.

Rowley Shoals, Timor Sea

Rowley Shoals *(top)* is located northwest of Australia. The atolls have 233 coral species and 688 species of fish.

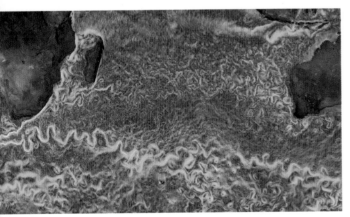

Arctic Ocean

The Terra satellite captured the Arctic Ocean *(middle)* in a partial solar eclipse, March 20, 2015.

Simulated Currents

This image of the Earth's liquid and frozen ocean system *(bottom)* is from a simulation conducted at NASA's Ames Research Center.

Image credit *(top):* NASA/JSC
Gateway to Astronaut Photography of Earth

Image credit *(middle):* NASA Goddard MODIS Rapid Response Team

Image credit *(bottom):* NASA/Ames Research Center

Fresh Water

Fresh water is contained in ice sheets, ice caps, glaciers, icebergs, bogs, ponds, lakes, rivers and streams, and aquifers. Approximately 97 percent of the Earth's water is salt water, leaving a little less than 3 percent fresh water.

Yamzho Yumco Lake, Tibet

Yamzho Yumco (Sacred Swan) Lake is high in the mountains of Tibet *(below)*. Its surface is about 14,570 feet (4,441 meters) above sea level.

Image credits: NASA/METI/AIST/Japan Space Systems, and U.S./Japan ASTER Science Team

Image credits: NASA, ESA, and ISS Crew Earth Observations

Lakes and Ponds

Lakes can be large or small, filled or dried up, salty or fresh, frozen or liquid, below or above ice, natural or constructed by humans. They are an important source of fresh water, especially away from rivers and the ocean. They are used as a water reserve for agriculture, human consumption, manufacturing, transportation, and recreation.

West of New South Wales, Australia, flooding on the Darling River *(top image, dark line)* forms ephemeral lakes. The community uses the lakes for water control and agriculture. The photograph was taken by the ISS Crew Earth Observations Experiment and the Image Science & Analysis Laboratory, Johnson Space Center.

Canada's Northwest Territories and Nunavut have numerous lakes and rivers that are used as seasonal ice roads in many arctic areas. This image *(left)* includes Amundsen Gulf, Great Bear Lake, and many other small lakes in the area.

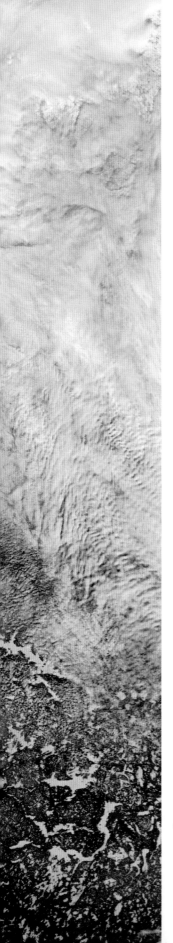

Image credits: NASA Earth Observatory images by Joshua Stevens, using MODIS data from LANCE/EOSDIS Rapid Response

The Susitna Glacier

Glaciers, like liquid rivers, have tributaries, although these are slow moving ice tributaries. In this false-color image that was made by recording infrared, red, and green wavelengths of Alaska's Susitna Glacier, vegetation is red.

Image credits: NASA/GSFC/METI/ERSDAC/JAROS, and U.S./Japan ASTER Science

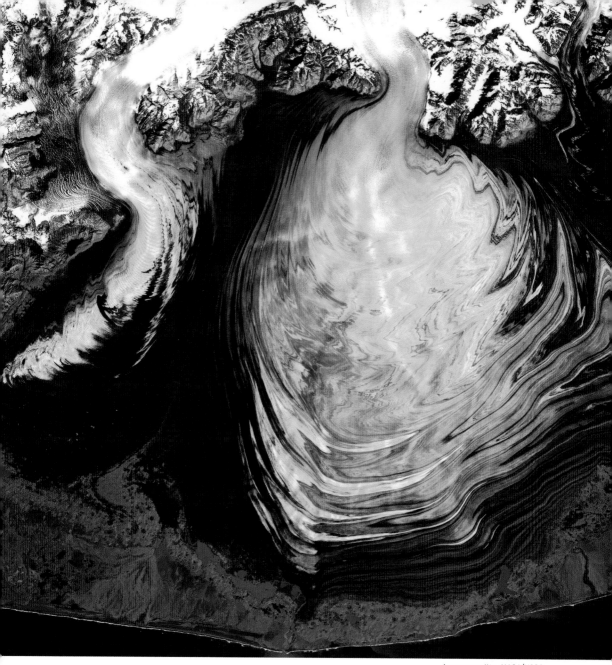

Image credits: NASA/USGS

Alaskan Glacier

This is an ice tongue that protrudes from the Malaspini Glacier, which is the largest glacier in Alaska. A tongue is formed when ice from the glacier moves quickly into a lake or ocean.

The false-color image, acquired by Landsat 7's Enhanced Thematic Mapper, 2000, was captured using infrared, near infrared, and green wavelengths.

Rivers

These images of the Mackenzie River, Canada, were captured using the Advanced Spaceborne Thermal Emission and Reflection Radiometer (ASTER) on August 4, 2005.

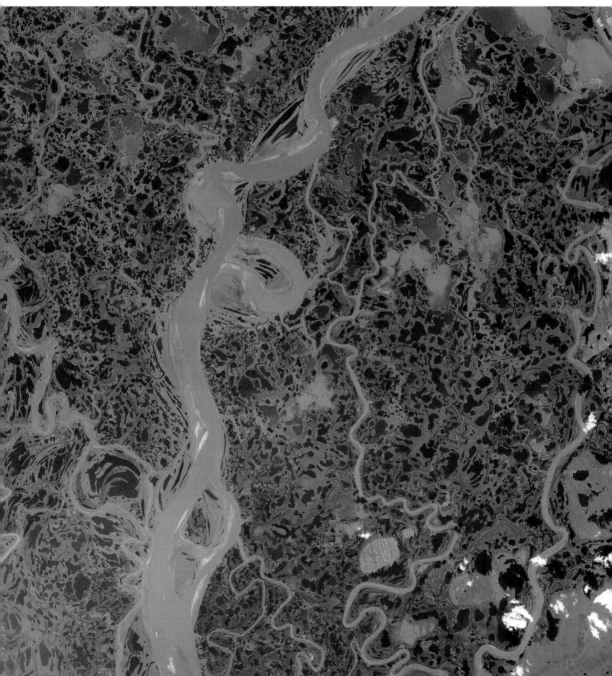

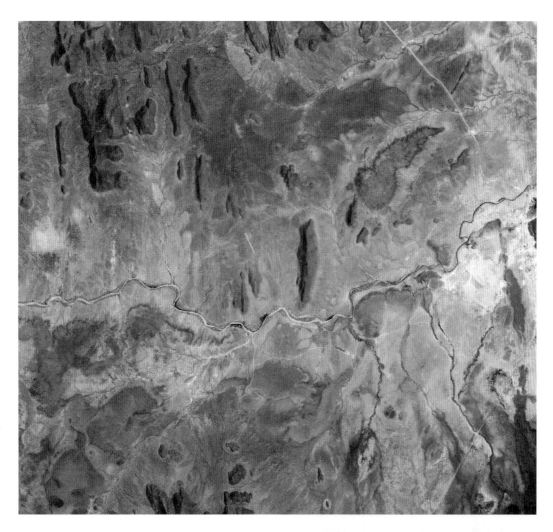

Not all rivers have water. Others have water only in the rainy season. This is the case in the region of Ewaso Nyiro River, Kenya *(top)*, where trees and other life are found only at the river's edges. Another river, the Chattahoochee River in the U.S. *(right)*, has enough water to facilitate thirteen dams and powers sixteen power-generating plants.

Image credit *(top):* NASA

Image credit *(bottom):* NASA Earth Observatory image created by Jesse Allen, using EO-1 ALI data provided courtesy of the NASA EO-1 team

Land

L and is often described in terms of landforms, which are features that occur naturally on the Earth's surface, even below the seas and oceans. Some of these landforms are basins, bays, canyons, deserts, hills, islands, mountains, peninsulas, plains, plateaus, ridges, valleys, and volcanoes.

Images of landforms taken from space are breathtaking and beautiful. Some reveal surprising textures, colors, and shapes that give us a fresh look at places we've seen in person or observed from the perspective of an Earth-based photograph.

Image credit :NASA

The Coast

The coast is where land and water come together. It's the interface between two very distinct environments. Weather and climate is impacted by this interface of two environmental systems that share a common boundary.

Night photographs of illuminated coastal areas *(below)* illustrate the considerable population density. Seen from space, the coastal areas of Cape Cod, Massachusetts *(facing page)* are fascinating and beautiful.

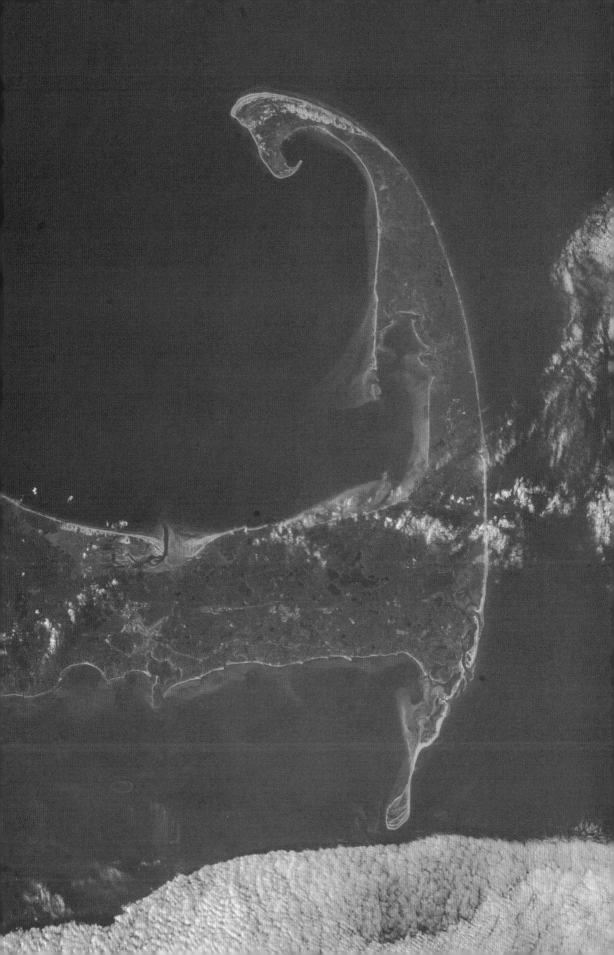

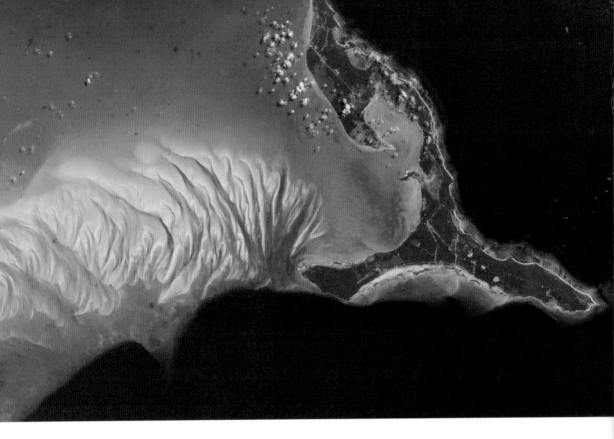

Islands

Vulnerable Islands

This image *(above)* taken in 2002, from the International Space Station, is of Eleuthera Island in the Bahamas.

This image *(below)*, also an International Space Station photograph, is of the Iles Eparses, located to the north and west of Madagascar. These islands are important for turtles and seabird nesting.

Many islands in the ocean are vulnerable to hurricanes and rising ocean levels. On September 6, 2017, Puerto Rico was enveloped by Hurricane Irma *(facing page, top right)*.

Although the island of Puerto Rico was devastated, it appears a little over one month later *(facing page, top left)* in serene weather.

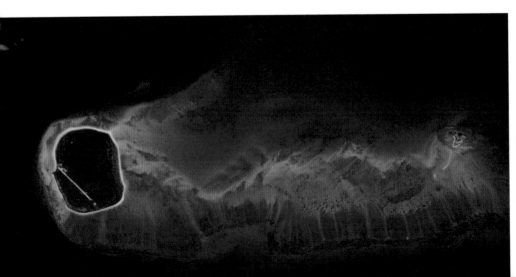

Image credit *(top):* NASA

Image credit *(bottom):* NASA

Bermuda

Bermuda *(bottom)* is a British Overseas Territory. It is actually not one island, but a chain of 181 islands.

Image credit : NASA

Islands of the Bahamas

This photograph was taken from the International Space Station by Expedition 52 Flight Engineer Randy Bresnik of NASA.

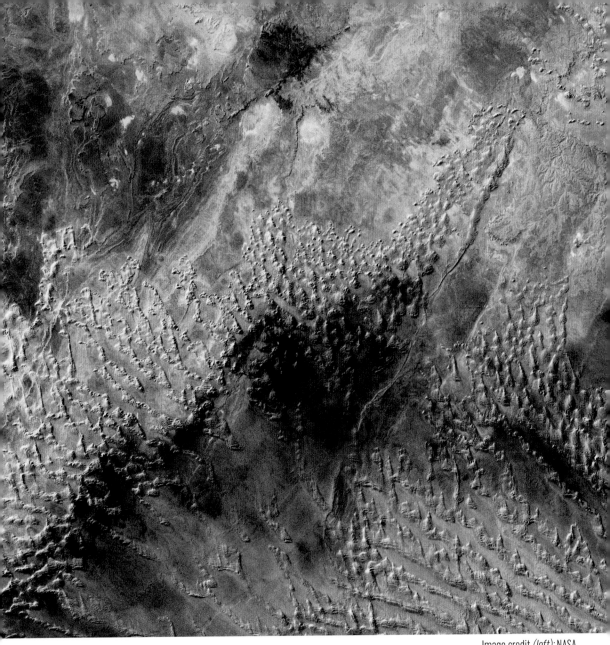

Gobi Desert

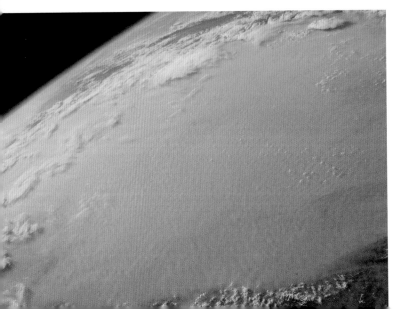

This image *(left)* of the Gobi Desert dust storms was taken by the crew of Expedition 43 on the International Space Station on April 29, 2015.

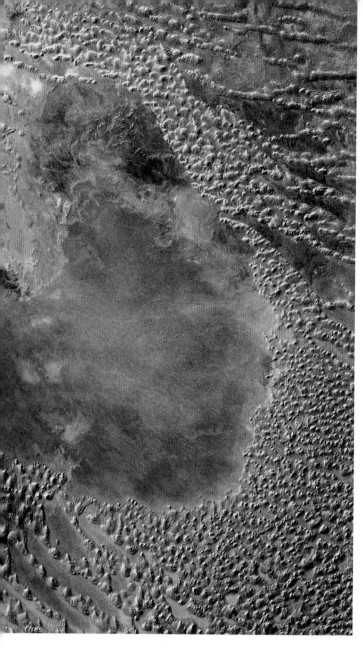

Sahara Desert

On October 3, 2016, this colorful image of the Sahara Desert *(left)* in western Libya was taken by the International Space Station Expedition 50 using the Sally Ride EarthKAM.

Rub' al Khali

Rub' al Khali *(below)*, one of the largest sand deserts, includes the southern Arabian Peninsula, part of Oman, United Arab Emirates, and Yemen. This image was acquired by ASTER aboard NASA's Terra Earth-orbiting satellite.

Image credit *(left):* Sally Ride EarthKAM

Image credits *(bottom):* NASA/GSFC/METI/ ERSDAC/JAROS, and U.S./Japan ASTER Science Team

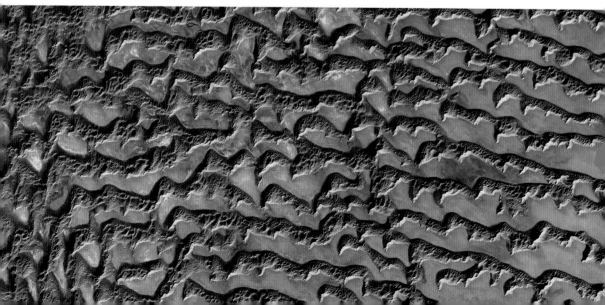

Mountains

European Space Agency (ESA) astronaut and Expedition 46 Flight Engineer Tim Peake took this photograph of the Alps *(top)* in winter 2015 from the International Space Station. The Alps are the highest mountains in Europe. At the time, the snow cover was light

The Aracar Volcano in the Andes Mountains *(middle)* shows a lava flow. This volcano was first know to be active in 1993 when a plume of steam or ash was seen in a nearby village. The volcano is located in northwestern Argentina, near the Chilean border.

The 2015 Winter Olympics were held in Sochi, Russia. This image *(bottom)* shows the Krasnaya Polyana Mountain Cluster where events were held. The images were made with ASTER using visible and near-infrared wavelengths and incorporating elevation data.

Image credits *(top):* ESA/NASA

Image credits *(middle):* NASA/JPL/UCSD/JSC

Image credits *(bottom):* NASA Earth Observatory image by Jesse Allen and Robert Simmon, using EO-1 ALI data from the NASA EO-1 team, archived on the USGS Earth Explorer

Tallest Mountain

The tallest mountain above sea level is Mount Everest, also known as Sagarmth and Chomolungma. Its summit lies between Tibet and Nepal. Everest is 29,029 feet (8,848 meters) high.

Not far below its summit, the sedimentary rock contains remnants of shallow-water marine organisms. This indicates that the mountain's summit was originally at sea level. Scientist say the mountain is still rising at about 5 millimeters per year.

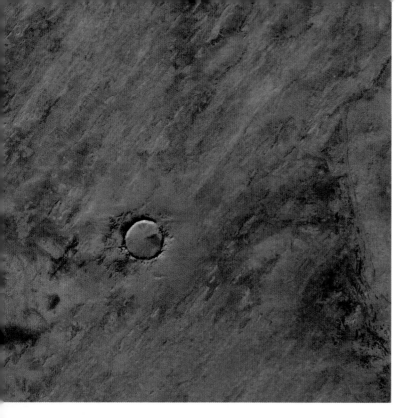

Craters

Craters can occur for a variety of causes. The several shown here are formed from meteorite impacts and volcanos.

Tenoumer Crater *(top)*, in the Sahara Desert, is 1.2 miles (1.9 kilometers) wide. The meteorite impact was on ancient rock and happened between 10,000 and 30,000 years old. The image was made with Advanced Spaceborne Thermal Emission and Reflection Radiometer (ASTER) in 2008.

Kebira Crater is a large impact crater located in the western desert of Sahara Desert and is probably over 100 million years ald. From the ground, the crater was speculated to exist due to findings of melted sand that turned to glass. After so much time, erosion had softened its shape, but it remained distinguishable from space— as seen in this 2001 Landsat-7 image *(bottom)*.

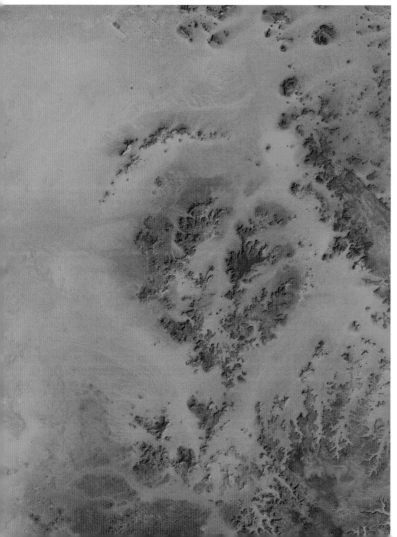

Image credit *(top):* Jesse Allen, using data provided courtesy of NASA/GSFC/METI/ERSDAC/JAROS, and U.S./Japan ASTER Science Team.

Image credit *(bottom):* Robert Simmon, based on Landsat-7 data provided by the UMD Global Land Cover Facility

The Wolfe Crater *(top)* seen in the upper left is an impact crater that is estimated to have been hit by a meteorite about 300,000 years ago. It is about 2,890 feet (880 meters) in diameter. The white dot in the middle is most likely from a gypsum deposit.

Crater Lake, Oregon *(bottom left)* is the deepest lake in the U.S. at an average depth of 1,148 feet (350 meters). About 7,700 years ago, the crater and lake were formed after the volcanic eruption of Mount Mazama, which is now dormant.

Another crater lake, Lonar Crater, India *(bottom right)*, was formed by a volcano between 35,000 and 50,000 years ago.

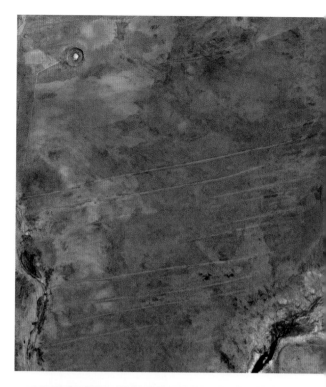

Image credit *(top):* Jesse Allen, based on data provided by the USGS Land Processes Distributed Active Archive Center

Image credit *(bottom right):* NASA image created by Jesse Allen, using data provided courtesy of NASA/GSFC/METI/ERSDAC/JAROS, and U.S./Japan ASTER Science Team

Image credits *(bottom left):* NASA/ISS Crew Earth Observations

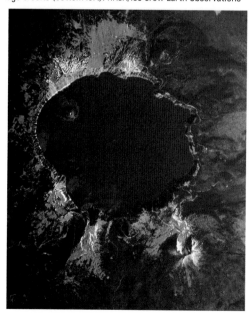

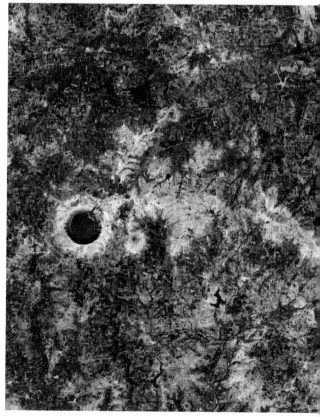

Life's Footprint

Human Activity

All life affects the Earth in some way. Examples of this—such as the plankton bloom in the North Sea *(page 38)* and the phytoplankton bloom in the Barents Sea *(page 123)*—are often revealed in images from space. However, it is the human footprint that is overwhelmingly evidenced in images from space.

The images on this page and the facing page are images taken at night that document the lights from urban areas. Human activity continues beyond daylight and into the night with the help of generated electricity. The shape of Italy *(top)* can be easily identified as the lights emphasize the peninsula's shape. In the image of Salt Lake City, Utah *(upper middle)*, a difference can be seen between the dense brighter lights of the city and dimmer lights of the suburban area. The cities of Cairo and Tel Aviv are on the coast *(lower middle)*, but much of this area is sparsely populated.

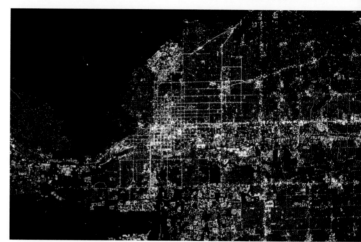

Image credit *(facing page):* NASA

Image credit *(top):* NASA, Expedition 49, ISS

Image credit *(upper middle):* NASA, Expedition 38, ISS

Image credit *(lower middle):* Expedition 49, ISS

Image credit *(bottom):* NASA

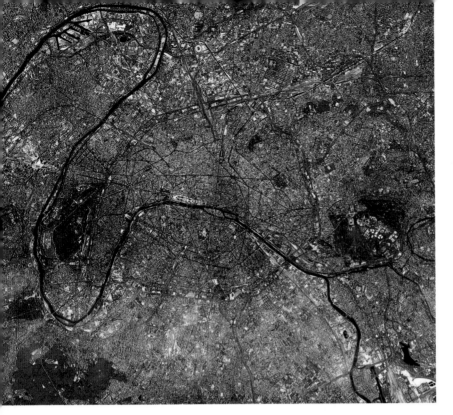

Cities and Towns

Over half of the Earth's population lives in cities. Satellite images and photographs from space of individual urban centers reflect their own unique geography, culture, history, land-use strategies, and technology. Housing, transportation, sanitation, utilities, and communication systems reveal their impact as well. Acquired with the ASTER instruments on the Terra satellite, here we see: Paris, France *(top)*, Las Vegas, Nevada *(bottom)*, and Beihai, China *(facing page)*.

Image credits *(top):* NASA/GSFC/ METI/ERSDAC/JAROS, and U.S./Japan ASTER Science Team

Image credit *(bottom):* NASA's Goddard Space Flight Center

Image credits: NASA/GSFC/METI/ERSDAC/JAROS,
and U.S./Japan ASTER Science Team

Farms and Agriculture

These images are a small sample of the many kinds of farming and agriculture found on Earth. Some practices are ancient, some are controversial, and some are state-of-the-art.

Westland in the Netherlands *(top left)* has taken farming to the utmost in innovation. Although a small country, its food production is second only to the U.S. yield.

Nile Delta fisheries *(bottom left)* contain hundreds of aquatic ponds. One of China's extensive fish farms *(facing page, top left)* was photographed from the International Space Station. Ginseng farms *(facing page, top right)* in northern China provide cover for these plants that need shade. These Kansas *(facing page, bottom)* crop circles are a result of irrigation systems.

Image credits *(top):* NASA/METI/AIST/ Japan Space Systems and U.S./Japan ASTER Science Team

Image credits *(bottom):* NASA/ GSFC/METI/ERSDAC/JAROS, and U.S./Japan ASTER Science Team

Image credits *(facing page, top left):* NASA and astronauts aboard the International Space Station

Image credit *(facing page, top right):* NASA

Image credits *(facing page, bottom):* NASA/GSFC/METI/ERSDAC/ JAROS, and U.S./Japan ASTER Science Team

Canals

Canals are artificial channels that carry water for the purposes of boat and ship transportation as well as irrigation or human consumption.

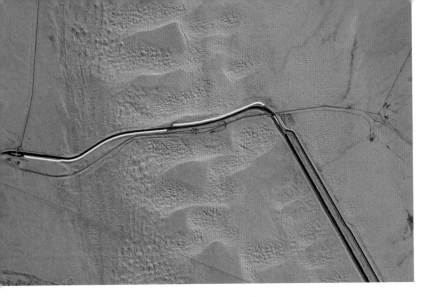

Image credit *(top):* NASA

The All-American Canal *(facing page, top)* is 80 miles long and runs east to west along the U.S.–Mexico border. Although it looks small and insignificant from the view of the astronaut photographers on the Expedition 18 on the International Space Station, over 500 people have died trying to cross it. The canal transports water for irrigation and eight hydroelectric power plants.

The city of Venice, Italy *(bottom)*, and its canals were originally marshy lagoons. Now they are twenty-one islands separated by canals and connected with bridges. This image was taken on February 14, 2017, Valentine's Day, by astronauts on the International Space Station with reference to the gondoliers and the lovers they carry on the canals through the city.

Image credit: ESA/NASA

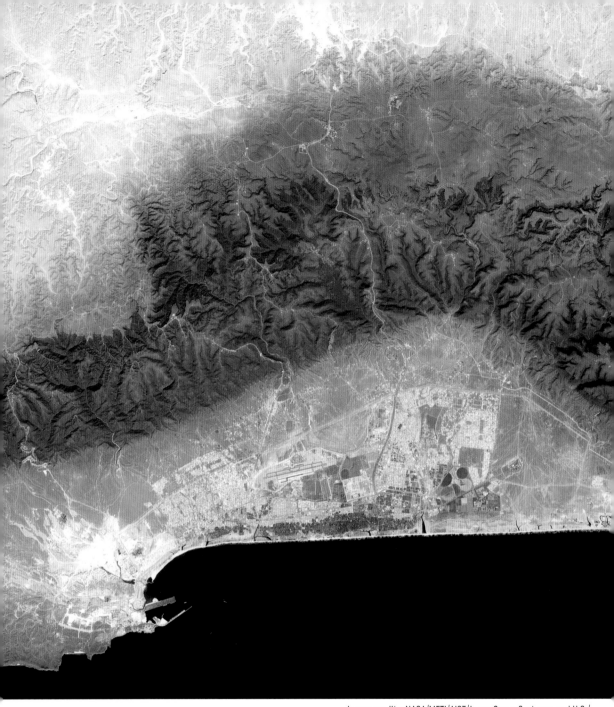

Image credits: NASA/METI/AIST/Japan Space Systems, and U.S./
Japan ASTER Science Team

Harbors and Ports

The Port of Oman *(top)*, is on the
Arabian Peninsula. Harbors, whether
formed naturally or constructed by

people, provide protection for ships.
Ports offer support facilities for the
loading of ships, the work of customs

Image credit *(top):* NASA

Image credits *(bottom):* NASA Earth Observatory images
created by Jesse Allen and Robert Simmon, using Landsat data
provided by the U.S. Geological Survey

officers, and others con-
ducting business activi-
ties. Many ports begin as
natural harbors and build
from there.

The image above
features the city lights
of Dubai, United Arab
Emirates. The port is the
largest and most popu-
lated city in the UAE.

Land reclamation for
Rotterdam, Netherlands,
can be seen underway in
this image *(right)*. Land
is taken from the sea.

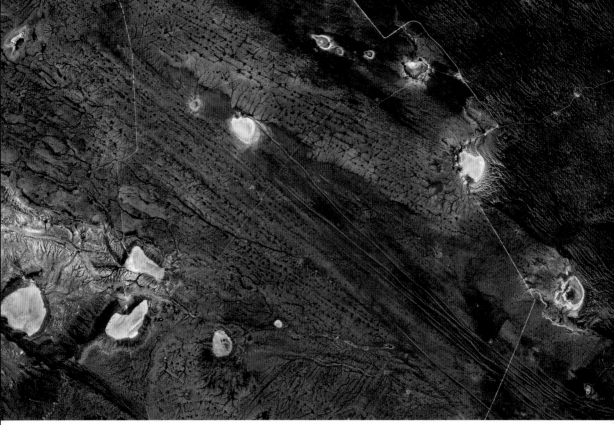

Roads and Highways

Image credit *(top):* modified Copernicus Sentinel data (2017), processed by ESA

Image credits *(bottom):* NASA/GSFC/METI/ERSDAC/JAROS, and U.S./Japan ASTER Science Team

Roads and highways are a necessary part of human life, whether we are traveling by land, water, or air. The courses of our travels look amazing from outer space; they show the beauty of our movements on Earth.

The image above is of lone roads crossing the Nabib Desert on the continent of Africa. Below, the bridge between Denmark and Sweden appears incomplete. However, Denmark's side of the road is actually a tunnel under the water.

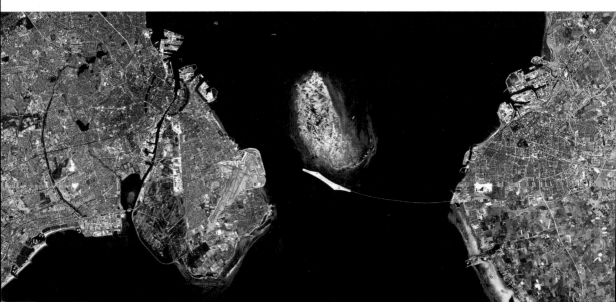

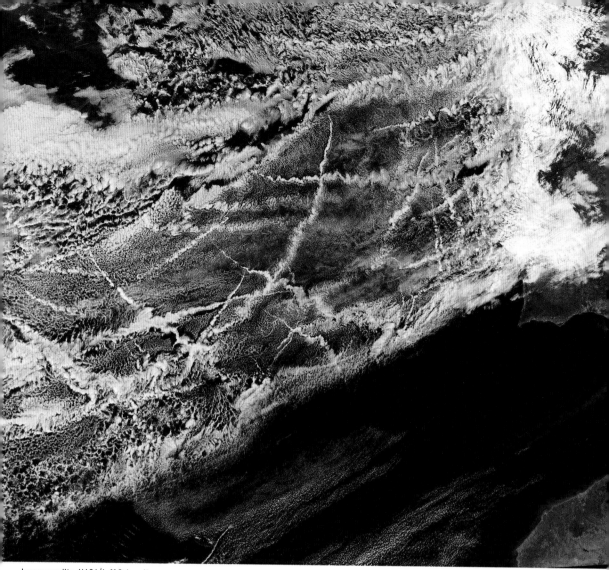

Image credits: NASA/Jeff Schmaltz,
LANCE/EOSDIS Rapid Response

Ship Tracks

Ship tracks form when water vapor
condenses on the pollution exhaust
from ships that crisscross the oceans.
The pollution particles act as seeds for
clouds to form around. This image
of the Atlantic Ocean show hundreds
and hundreds of miles of ship tracks.
Similar to contrails of jets in the sky,
the narrow ends are newer and the
wider, diffused end is older.

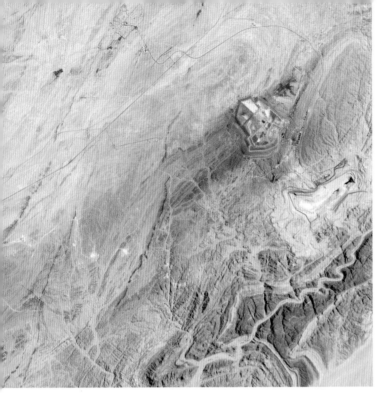

Mining

Mining can reshape the surface of the Earth, as can be seen in each of these images. Even in spite of reclamation, the crust can remain scarred. Ground water can also be contaminated.

The Rössing Mine *(top)*, seen here on the middle right of the image, is the oldest and third-largest producer of uranium. It's located in the Namib Desert.

Botswana's diamond mines *(bottom)* produce billions of dollars a year. There are four open pits with concentric roads that wind deeper and deeper.

This astronaut photographer's image *(facing page, top)* from the International Space Station, shows a coal mine in Germany. The coal is used to fuel a power station.

Cananea copper mine *(facing page, bottom left)*, in Sonora, Mexico, has deposits of copper and gold. The blue is an indication of those rich minerals.

Sunrise Dam Gold Mine *(facing page, bottom right)* is in Western Australia. Gold was discovered here in 1988 and the mine has quickly stepped up its development.

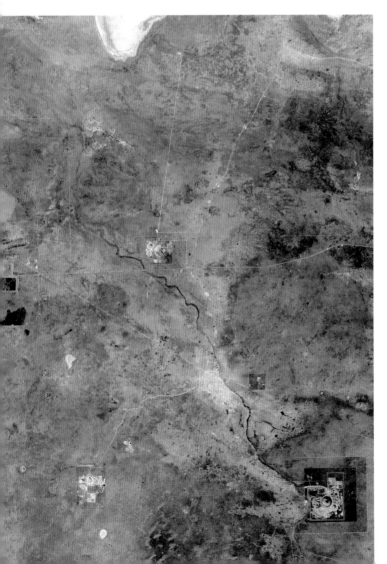

Image credits *(facing page, top):* NASA Earth Observatory image by Jesse Allen and Robert Simmon, using EO-1 ALI data

Image credit *(facing page, bottom):* the ISS Crew Earth Observations experiment

Image credit *(top):* NASA/JSC Gateway to Astronaut Photography of Earth

Image credit *(bottom, left):* NASA Earth Observatory image created by Jesse Allen, using EO-1 ALI data

Image credit *(bottom, right):* NASA Earth Observatory image created by Jesse Allen, using EO-1 ALI data

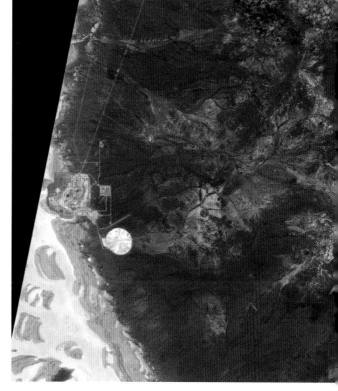

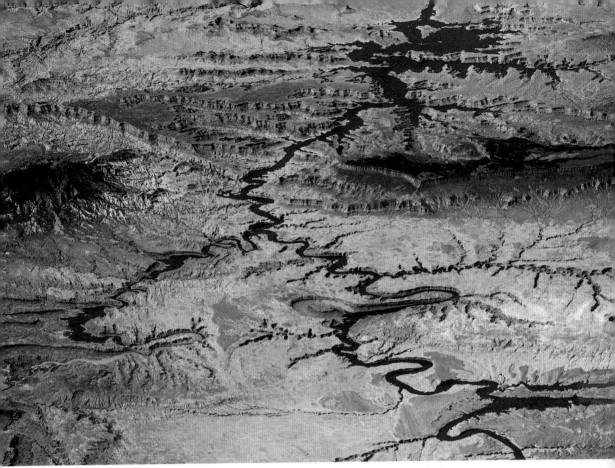

Image credit *(top):* NASA

Image credit *(facing page):* NASA Earth Observatory, created by Jesse Allen, using EO-1 ALI data provided courtesy of the NASA EO-1 team

Dams and Hydro Works

Dams and their reservoirs are used for recreation, flood control, hydro-electric power, irrigation, household water, and more. Lake Powell *(top)* was made by damming the Colorado River. The reservoir is thin and long as it follows the contour of Glen Canyon.

A large amount of ice is in the Kyiv Reservoir *(facing page)* on the Dnieper River, which is first in a series of dams on the river. The dam is about 50 miles (80 kilometers) southeast of the Chernobyl nuclear plant where a meltdown disaster occurred in 1986.

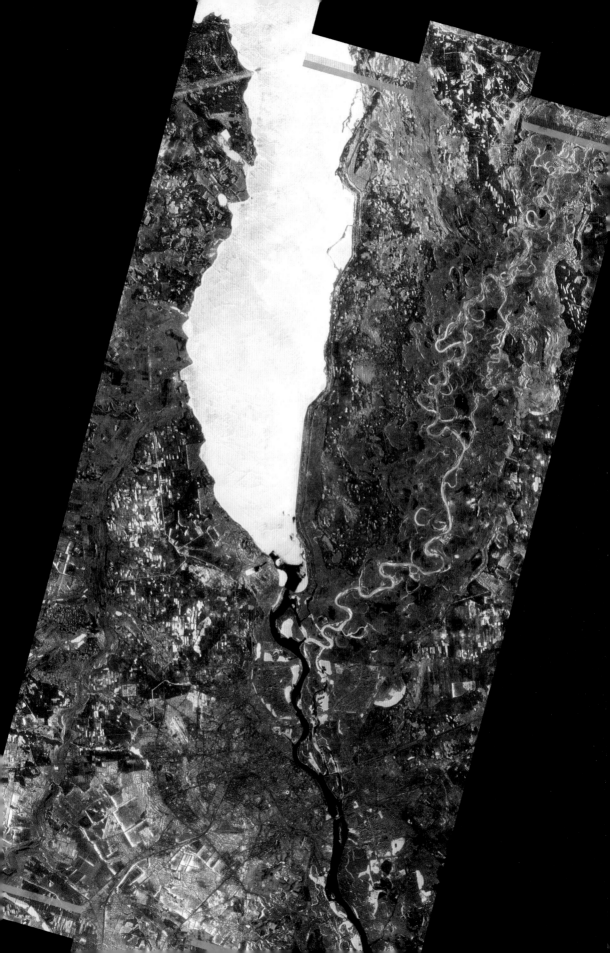

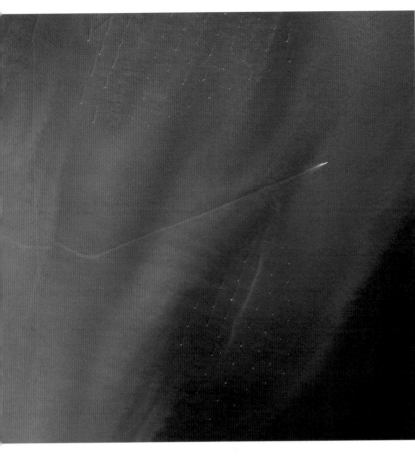

Green Energy

Green energy often takes up a large area to gather power. Windmills, mirrors, and photovoltaic cells need acres and acres to create the renewable clean energy.

This image *(left)* shows tiny white dots that are wind turbines set up in the North Sea. Each turbine also creates a wake or plume of sediment.

Image credit *(top, bottom, and facing page):* NASA image created by Jesse Allen, using data provided courtesy of NASA/GSFC/METI/ERSDAC/JAROS, and U.S./Japan ASTER Science Team

This image of farms *(facing page, bottom)* field near Seville, Spain, looks like a crazy quilt. Nearby, a collection of large mirrors reflect the sunlight onto a steam generator that is used to turn generator turbines.

At the time it was photographed in 2015, this photovoltaic power plant *(above)* was the largest in the world.

Dynamic Earth

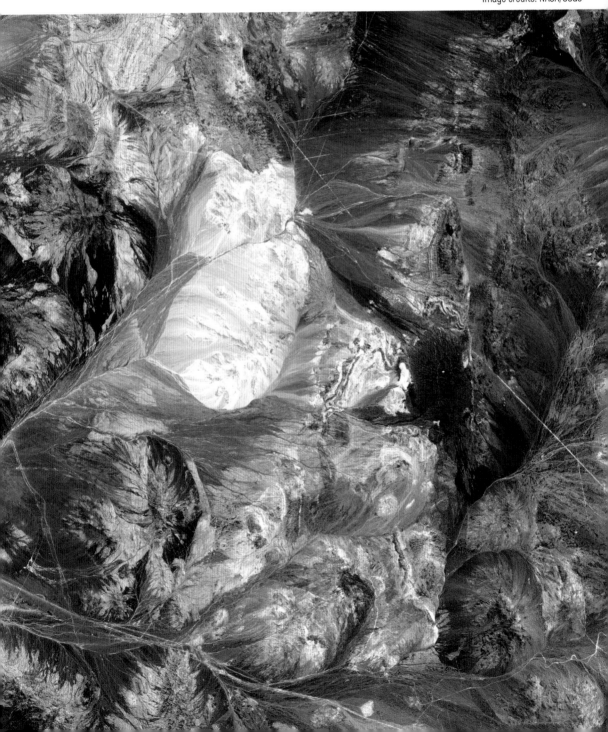

The Earth is always changing. It reacts to the power of the forces put upon it such as the gravitational pull of the Moon and its sun. There are also forces within the Earth, such as volcanos and earthquakes. And there are forces on the surface, such as weather and human activities. Although the Earth is resilient, change is inevitable. Forces on Earth are dynamic and compel change.

Reluctant Change

A desert may seem unhurried and reluctant to change. However, the Atacama Desert in Chile *(facing page)* looks quite dynamic from space. With its sleeping volcano, it's only a matter of time before its force takes hold.

Image credits: NASA/GSFC/METI/ERSDAC/JAROS - NASA Earth Observatory

Pyroclastic Flow

This false-color image *(above)* shows a pyroclastic flow and a plume of ash near Mount Merapi on Java of Indonesia.

Volcanic Eruptions

This is Sarychev Volcano *(top)*, on Russia's Kuril Islands, from the vantage point of the International Space Station, in 2009. The image of the volcano's eruption was caught at its early stage.

Satellites recorded small but detectable changes in the Heard Island lava flow on Mawson Peak in Australia *(left)*, in 2012, suggesting volcanic activity.

Cleveland Volcano in the Aleutian Islands *(top right)* was photographed in 2008 from the International Space Station. The plume of ash was reported from the International Space Station (ISS) by Flight Engineer Jeff Williams.

This volcano on the Kamchatka Peninsula, Russia, *(bottom)* was also photographed by the International Space Station crew in 2013. The plume is most often steam, volcanic gases, and ash.

Image credit: NASA

Image credit: NASA

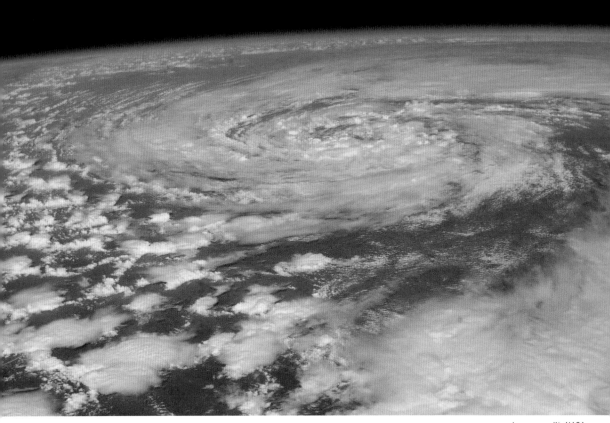

Image credit: NASA

Storms

Storms are violent, life threatening, and destructive. A major reason for improving weather forecasting is for storm preparation, to eliminate loss of life, and to reduce property damage. However, from an astronaut's perspective, storms can be beautiful. They are also necessary. They bring the rain and create fresh water. They stir things up. Although storms are devastating, they stimulate growth by clearing away the dead and decaying, then circulating the nutrients.

This Geocolor image *(facing page, top)* from NOAA's GOES-16 Satellite shows an unusually strong winter storm in the winter of 2018 in the northeastern United States. The Geostationary Operational Environmental Satellite (GOES) is a collaboration between NOAA and NASA. The GOES-16 satellite is the first to provide continuous imagery and data for monitoring weather. Infrared is used at night.

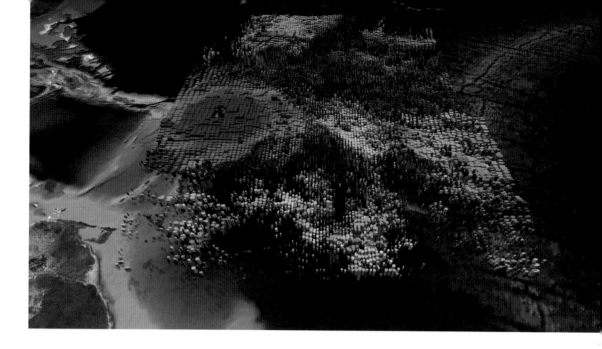

Visualizing the Storm

Hurricane Irma's clouds are typical of a tropical storm *(below)*. The data for this image came from the AIRS instrument on NASA's Aqua satellite.

Each cylinder *(above)* indicates the cloud thickness. The colors indicate the temperature at the tops of the cloud area. Because storms are dynamic, an animated version of this image is also used.

Image credit *(top):* NASA/JPL-Caltech

Image credit *(bottom):* NOAA

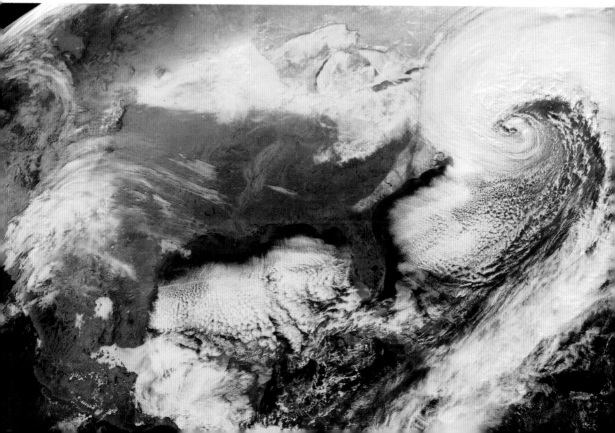

Flooding

Flooding results from rain, storms, tsunamis, or water releasing from natural or human-constructed dams, creating water levels that are above normal for the area. The water can be destructive, and it displaces human, animal, and plant life. It can change the environment temporarily or per-manently. It can also rejuvenate life in certain habitats.

Compare the top image of Lake Enriquillo, in the Dominican Re-public, with the bottom image after flooding. Much of the surround-ing land is submerged and an island has been created.

Drought

Juxtaposing images of water bodies experiencing drought with images of their normal water levels, dramatically reveals an altered landscape.

The image of Lake Mead *(top)* in the U.S. was taken in 2000 by Landsat 7. The bottom image was taken in 2015 by the ASTER instrument on the Terra satellite. The lake's size and depth is reduced, islands are larger or have become part of the main land, and the lake's contour has changed.

Image credits *(facing page, top and bottom)*: NASA/METI/AIST/Japan Space Systems, and U.S./Japan ASTER Science Team

Image credits *(top and bottom)*: NASA/METI/AIST/Japan Space Systems, and U.S./Japan ASTER Science Team

Landslides

Landslides happen when an unstable mass of rock, earth, and/or debris slides down a mountain or hill. They can happen anywhere there is a change in elevation. Some are large.

Some are small. Either can be deadly. There is very little that can be done to prevent landslides. Loss of life and property can be minimized by locating buildings in safer, more stable areas that are less prone to slides.

The top image (acquired with ASTER) shows a Tibetan landslide, on July 17, 2016, that killed nine people. Because of the amount of ice, this is considered an avalanche. However, since this area has glaciers, which contain rocks and debris as well, both terms seem to be used in descriptions of the event.

The bottom image (also ASTER) shows the Blackhawk landslide, in Lucern Valley, California, which is estimated to have happened around 18,000 years ago. The slide is about five miles long and two miles wide, one of the largest in North America.

Image credits *(top and bottom)*: NASA/METI/AIST/Japan Space Systems, and U.S./Japan ASTER Science Team

Image credits: NASA/METI/AIST/Japan Space Systems, and U.S./Japan ASTER Science Team

Earthquakes

The image shows the earthquake-prone area on the border of Iran and Iraq near Halabjah, Iraq. On November 12, 2017, the area experienced a magnitude 7.3 quake with casualties and extensive damage. The yellow star shows the epicenter. Red areas denote planted fields, and pale red indicates shrubs and trees. Dark gray areas indicate earlier brush fires. Tan and lighter gray represent various types of rocks.

Fire

Fires can be either controlled burns (for land and agriculture management) or out-of-control events (started by people or lightning strikes). In either circumstance, the Advanced Spaceborne Thermal Emission and Reflection Radiometer (ASTER) instrument on NASA's Terra satellite help firefighters manages these fires.

The top image shows the Thomas Fire, the largest wildfire in California's recorded history. It burned 282,000 acres0 destroying 1,063 structures. It was spread by high winds, high temperatures, and low humidity.

The close-up of the Thomas Fire *(bottom left)* shows where new areas were ignited as a result of the wind blowing embers as far as two miles away into dry brush.

Image credits *(top):* NASA/METI/AIST/Japan Space Systems, and U.S./Japan ASTER Science Team

Image credits *(bottom):* NASA image courtesy Jeff Schmaltz LANCE/EOSDIS MODIS Rapid Response Team, GSFC

Agricultural Fires

Agricultural fires on the Island of Madagascar *(right)* are deliberate burns that clear the fields of debris and add nutrients to the soil. World-wide, this slash-and-burn farming technique accounts for considerable smoke that has substantial polluting effect on the atmosphere.

Image credits *(bottom):* NASA image courtesy Jeff Schmaltz LANCE/EOSDIS MODIS Rapid Response Team, GSFC

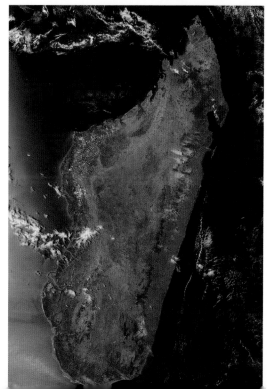

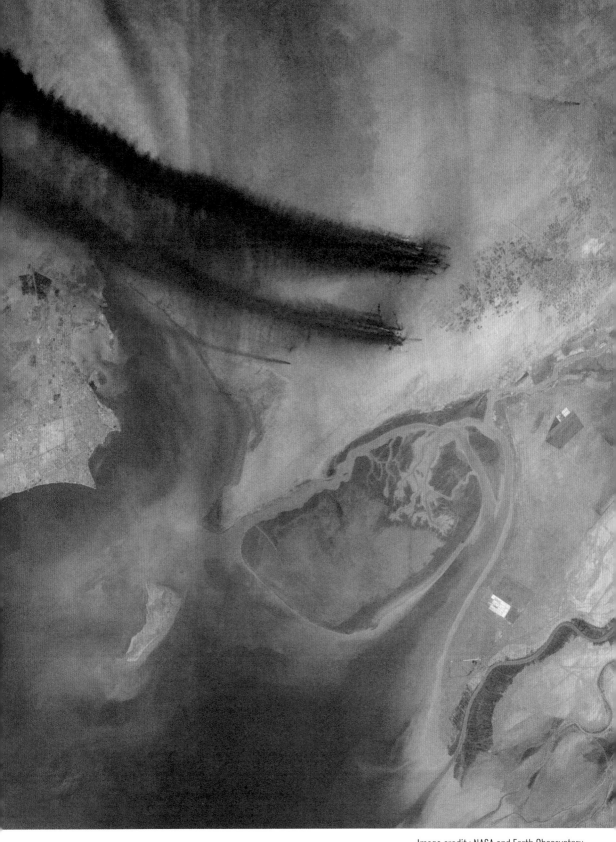

Image credit : NASA and Earth Observatory

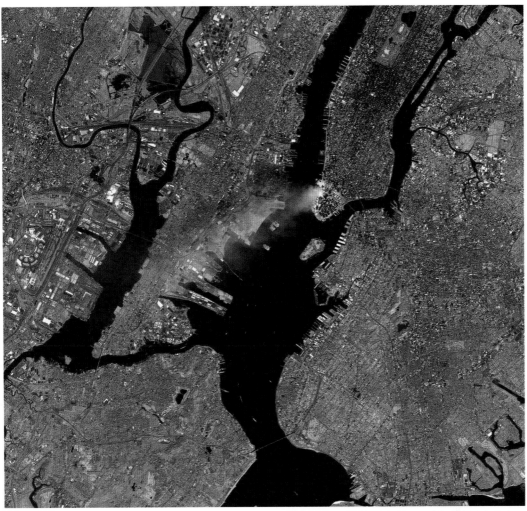

Image credit : USGS Landsat 7 team, at the EROS Data Center

Scorched Earth

Terrorism

Scorched earth describes a military policy where something of use (for example, crops, water, or oil) is destroyed. In 1999, Iraqi military forces set fire to 750 oil wells across Kuwait *(facing page)*. Besides reduced air quality while the fires burned, Kuwait may have clean up and long-term groundwater consequences for their already limited supply.

Terrorism is devastating to both life and the Earth itself. The top photo shows the aftermath of the World Trade Center attack as seen from the Landsat 7 satellite on September 11, 2001 at roughly 11:30AM.

Tsunamis and Waves

Tsunamis, long waves caused by an earthquake, are deep waves and do not gain height until approaching shallow coastal water. Consequently, tsunamis are difficult to see and track across the oceans. This image *(above)*, acquired with the Advanced Land Imager (ALI) on the Earth Observing 1 satellite on March 6, 2007, shows two kinds of waves: shallow surface waves and deep ocean waves. Scientists studying the deep waves are gaining an understanding of how the waves change as they move across the ocean and are affected by the underwater terrain.

Vulnerable Coasts

Tsunamis and storms have an impact on coastal communities, especially islands like Sri Lanka *(facing page)*. The red indicates areas that are vulnerable to flooding linked with storm surges, rising sea level, and tsunamis.

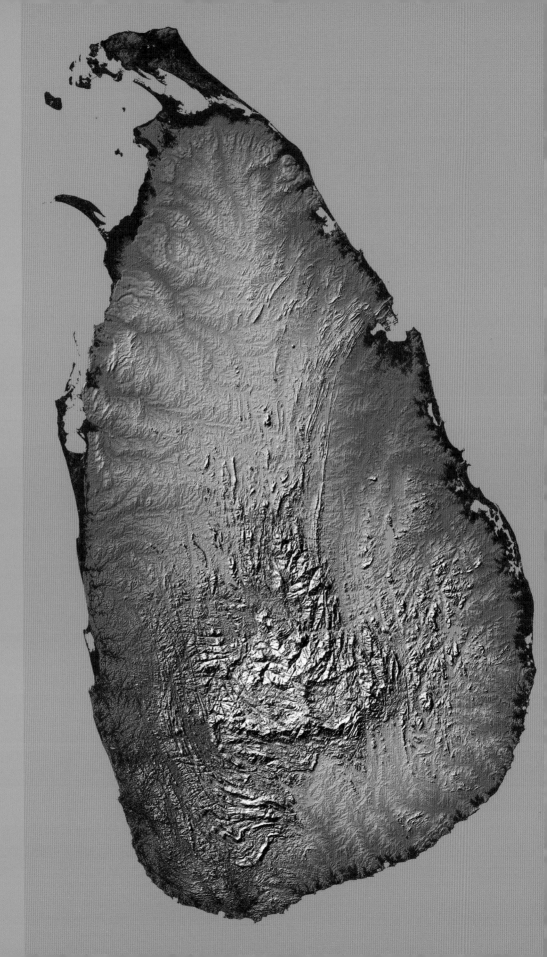

Earth from Far Way

From the Moon

This is how our home planet Earth *(below)* looks as seen from the Moon. Apollo 8, the first manned mission to the Moon, took this photograph.

From Mars

This composite image *(facing page, top)* of the Earth and its moon was acquired on November 20, 2016, by the HiRISE camera on NASA's Mars Reconnaissance Orbiter. The Earth and its moon's corresponding sizes and distance are correct in proportion.

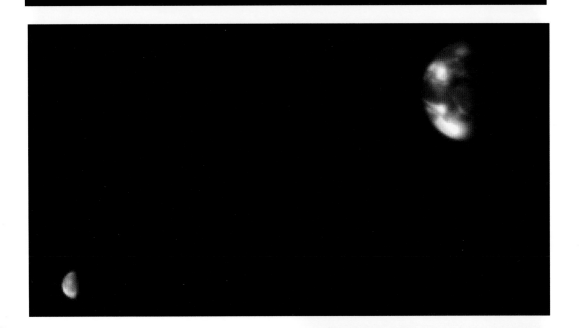

NASA's Mars Odyssey spacecraft took this photograph *(upper middle)* of Earth and the Moon in 2001 from over two million miles away.

From Saturn

The two bottom images were taken from NASA's Cassini spacecraft in July 2013. They show how tiny Earth looks from the perspective of the distant planet Saturn.

Image credits *(bottom):* NASA/JPL-Caltech/Univ. of Arizona

Image credits *(facing page, top):* NASA/JPL-Caltech/Univ. of Arizona

Image credits *(upper middle):* NASA/JPL-Caltech/Univ. of Arizona

Image credits *(lower middle):* NASA/JPL-Caltech/Space Science Institute

Image credits *(bottom):* NASA/JPL-Caltech/Space Science Institute

Observing Our Changing Earth

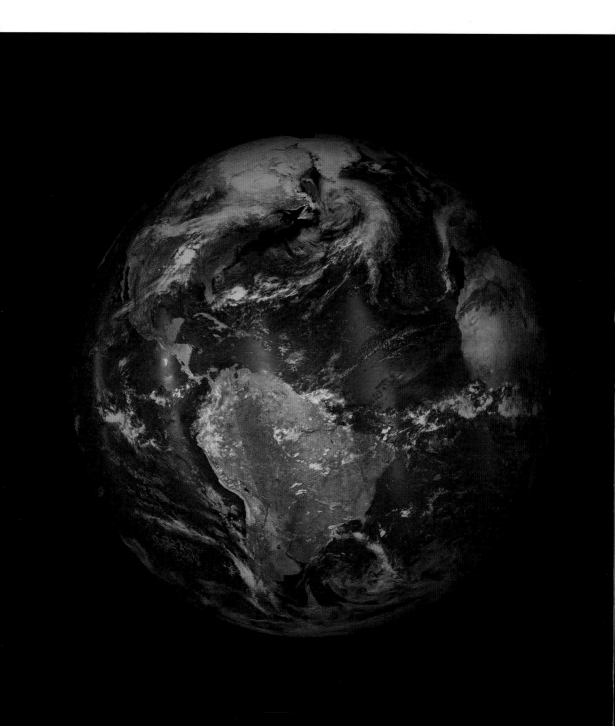

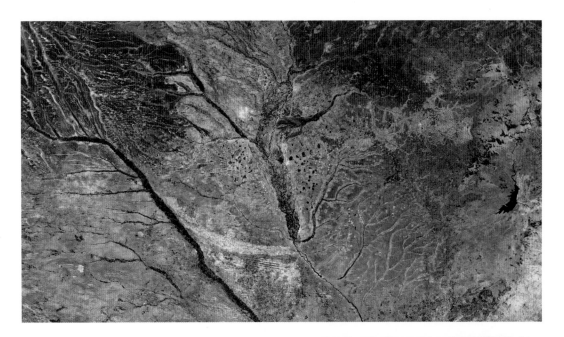

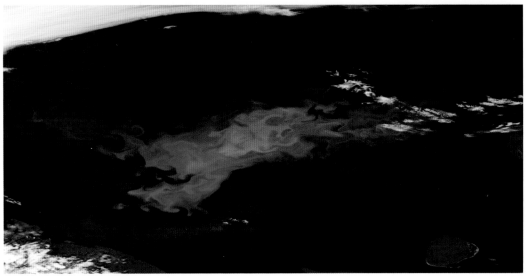

Watch the Earth

Image credits *(top):* NOAA National Environmental
Satellite, Data,and Information Service (NESDIS)

Image credits *(bottom):* NOAA National Environmental
Satellite, Data,and Information Service (NESDIS)

Image credits *(facing page):* NOAA National Environmental
Satellite, Data,and Information Service (NESDIS)

To be good stewards of our planet,
we will need to witness it with our
eyes and our instruments—and then
record, interpret, compare, anticipate,
educate, innovate, adapt, and enjoy.

Index

Other Books in This Series

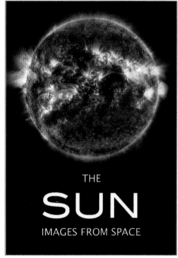

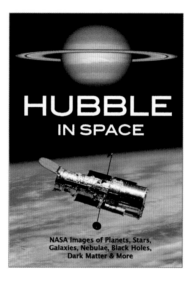

The Sun

Take a journey to the very center of our solar system and explore the awesome beauty of the star that fuels life on earth. Through fascinating descriptions and incredible color photographs from NASA and other space exploration organizations, you'll learn to see the sun in a whole new light. *$24.95 list, 7x10, 128p, 180 color images, index, ISBN 978-1-68203-340-1.*

Hubble in Space

Images from NASA's Hubble Space Telescope show the solar system, Milky Way, galaxies, and the far reaches of the universe up close like never before. See amazing details and explore the immense content of the universe. *$24.95 list, 7x10, 128p, 180 color images, index, ISBN 978-1-68203-300-5.*

The Moon

Take a virtual trip to the moon to examine its mysterious beauty, compelling terrain, and scientific significance. Images taken from satellites, space stations, massive telescopes, and NASA missions reveal incredible new stories to inspire deeper understanding. *$24.95 list, 7x10, 128p, 180 color images, index, ISBN 978-1-68203-368-5.*